UNDERSTANDING GLOBAL CRISES

Understanding Global Crises is an innovative and interdisciplinary text that investigates the key contemporary economic, social, and environmental crises and demonstrates their deep interconnection.

Contributing to the discussion of large-scale crises, this book provides a conceptual framework to understand the current global landscape. Essential cascading crises topics, such as economic collapse, climate change, racial injustice, domestic violence, and epistemic oppression, are explored in order to equip readers with the clarity to understand global crises, assess policy interventions, and analyze social responses. To achieve future resilience, the book shows that society must recognize various forms of inequality and make policy changes.

Each chapter showcases an international case study, covering real-life examples of topics such as climate disinformation, vaccine distribution disparities, environmental racism, and socioeconomic deprivation. Other features of the book include key terms, suggested further reading, and discussion questions, as well as online supplements comprising PowerPoint slides and an instructor's guide. *Understanding Global Crises* will be a valuable text to support courses in economics, environmental studies, political science, public health, and social policy.

Thomas R. Sadler is Professor of Economics at Western Illinois University. He teaches courses on Environmental Economics, Energy Economics, Pandemic Economics, the Global Economic Environment, and the Chicago Economy. His research focuses on environmental policy, energy economics, professional sports leagues, high-performance organizations, and neighborhood effects.

UNDERSTANDING GLOBAL CRISES

From Covid to Climate Change and Economic Collapse

Thomas R. Sadler

Routledge
Taylor & Francis Group

LONDON AND NEW YORK

Designed cover image: sanjeri / © Getty Images

First published 2023
by Routledge
4 Park Square, Milton Park, Abingdon, Oxon OX14 4RN

and by Routledge
605 Third Avenue, New York, NY 10158

Routledge is an imprint of the Taylor & Francis Group, an informa business

British Library Cataloguing-in-Publication Data
A catalogue record for this book is available from the British Library

Library of Congress Cataloging-in-Publication Data
Names: Sadler, Thomas R., author.
Title: Understanding global crises: from Covid to climate change and
economic collapse / Thomas R. Sadler.
Description: New York, NY: Routledge, 2023. | Includes
bibliographical references and index. |
Identifiers: LCCN 2022029140 | ISBN 9781032315058 (hardback) |
ISBN 9781032315027 (paperback) | ISBN 9781003310075 (ebook) |
ISBN 9781032378480 (ebook other)
Subjects: LCSH: Economic development. | Regional disparities. |
Environmental policy. | COVID-19 Pandemic, 2020—Influence.
Classification: LCC HD82 .S263 2023 | DDC 338.9—dc23/eng/20220825
LC record available at https://lccn.loc.gov/2022029140

ISBN: 978-1-032-31505-8 (hbk)
ISBN: 978-1-032-31502-7 (pbk)
ISBN: 978-1-003-31007-5 (ebk)
ISBN: 978-1-032-37848-0 (eBook+)

DOI: 10.4324/9781003310075

Typeset in Bembo
by codeMantra

Access the Support Material: www.routledge.com/9781032315027

CONTENTS

FIGURES

TABLES

CASE STUDIES

PREFACE

While most people living through the coronavirus pandemic experienced health and/or economic hardship, the period of time revealed how a series of cascading crises may unfold. The pandemic led to rising levels of morbidity and mortality. The economic collapse decreased both economic activity and employment. The unintended consequences of shutdown policies that slowed the spread of the novel coronavirus included several forms of social instability, including racial injustice, domestic violence, and epistemic oppression. At the same time, the climate catastrophe loomed over the health, economic, and social forms of instability, existing as the world's most important environmental problem. Using a multi-disciplinary perspective, *Understanding Global Crises* investigates the implications of these interconnected problems, arguing that they disproportionately impact the most vulnerable members of society. As the examples and case studies in the book demonstrate, interventions that intend to slow the spread of disease or boost economic activity should also consider the unintended consequences of policy design.

ACKNOWLEDGMENTS

I would like to thank Routledge for publishing this book. At every stage of the writing process, members of the editorial staff were professional, helpful, and responsive. I would like to thank the first economics editor of the project, Natalie Tomlinson, and the second economics editor, Michelle Gallagher, for their encouragement and help throughout the process. The editorial assistants, Helena Parkinson and Chloe Herbert, provided guidance, answered questions, and helped to bring the project to completion. At Western Illinois University (WIU), I have colleagues who engage in research, provide seminars, and establish an environment for scholarly work, including Tara Feld, Alla Melkumian, Jessica Lin, Shankar Ghimire, J. Jobu Babin, William Polley, Warren Jones, Tej Kaul, Kasing Man, Anna Valeva, Rong Zheng, Haritima Chauhan, and Feng Liu. I appreciate their ongoing support. I would like to thank the WIU Foundation and Office of Sponsored Projects for a summer stipend that contributed to the project. I have additional colleagues who provide the opportunity for coauthorship, including John Tomer, William Koch, Shane Sanders, Bhavneet Walia, and Justin Ehrlich. The mentors throughout my career have provided invaluable support and guidance, including William Kleiner, David Loschky, and Robert Bohm. For decades, Fred Ebeid has been a mentor and friend. Thank you for everything. My Mundelius cousins, Tony, Michael, John, and Matthew, provide the opportunity for both extended family and adventure, especially skiing, ice blocking, and stadium visits. My Illinois aunt and uncle, Chris Kaufman and Mark Austill, and California aunt and uncle, Patricia and Fred Mundelius, have always provided love and support. Thank you very much. I would like to express special thanks to my sister, Laura Sadler, and Rick Hazen for their special place in the family. My parents, Judith K. Sadler and Charles G. Sadler, establish a loving family environment that values education and the learning process. I can't thank you enough. My children, Maya and Mathew, are the joys of my life and the reasons I continue with this scholarship. Thank you for the fun times, summer trips, and innings in the leftfield bleachers at Wrigley Field. To my wife, Holly Stovall: I love you very much. Thank you for the daily conversations about all the topics in this book. I couldn't have written it without you.

AN OVERVIEW OF THE BOOK

The book is organized in three parts. Chapter 1 introduces the topic of global crises. Part I then investigates health, economic, and climate crises, including the coronavirus pandemic (Chapter 2), economic collapse (Chapter 3), and climate catastrophe (Chapter 4). Part II considers social instability, including racial injustice (Chapter 5), domestic and family violence (Chapter 6), and epistemic injustice (Chapter 7). In Part II, instructors interested in specific topics may cover the chapters in any order. Part III addresses the concept of building future resilience, including progress or collapse (Chapter 8) and resilience and vulnerability (Chapter 9).

LEARNING OBJECTIVES

Each chapter lists learning objectives. By linking to chapter outlines and material, the learning objectives provide continuity. But the learning objectives also link to chapter takeaways at the end of each chapter. With this framework, students will understand chapter structure, the progression of ideas, and important points for review.

EBOOK+ FEATURES

An added element to the book is eBook+ features, which include figures with color and pop-up definitions. The pop-up definitions throughout the text correspond to a list of key terms at the end of each chapter.

1

GLOBAL CRISES

An introduction

Chapter learning objectives

After reading this chapter, you will be able to:

LO1   Evaluate the era of cascading crises.
LO2   Characterize crises and their outcomes.
LO3   Indicate the potential for chain reactions from network connections.
LO4   Argue that unintended lockdown effects include economic and social instability.
LO5   Recognize that intersectional factors create interdependent forms of disadvantage.
LO6   Emphasize the ongoing problems of poverty, discrimination, and inequality.
LO7   Use a systems approach to address statistical patterns in nondiscriminatory ways.
LO8   Identify economic, political, and social problems in an age of discord.

Chapter outline

Cascading effects
Characteristics of crises
Networks
Unintended consequences
Intersectionality
Poverty, discrimination, and inequality
Systems approach
Age of discord
Summary

DOI: 10.4324/9781003310075-1

Cascading effects

Large-scale and interconnected **crises**—times of intense difficulty, destruction, and danger—characterize the current age. The pandemic, economic volatility, climate change, and social instability exist as features of the contemporary landscape. They are global in scale. They destabilize societies, reducing the ability of governing authorities, market mechanisms, and social institutions to implement resilient solutions. As Ed Yong (2020) argues in *The Atlantic*, "We have no choice, though, but to grapple with them. It is now abundantly clear what happens when global disasters collide with historical negligence. . . . Recovery is possible, but it demands radical introspection." The stakes are high. As a result, *Understanding Global Crises* provides introspection for these interconnected problems. The analysis is important, because recovery initiates the potential for creative, enduring, and regenerative outcomes. Recovery also provides an incentive for transformative change. But lingering economic, environmental, health, and social problems increase the potential for decline, the weakening of institutions, processes, and systems.

As a global **shock**—a sudden large-scale event—the coronavirus pandemic increased morbidity and mortality, burdened healthcare systems, ravaged the global economy, increased unemployment, devastated global supply chains, and overwhelmed social services. In contrast to earlier coronavirus outbreaks—Severe Acute Respiratory Syndrome (SARS-Cov-1) in 2003 and Middle East Respiratory Syndrome (MERS) in 2012—the SARS-CoV-2 virus in 2020 had a higher rate of transmission. It also spread in human transmission **networks**, the global mechanisms of interconnection. The disease associated with the novel coronavirus, Covid-19, had "no known preexisting immunities, (was) spread by people that (did) not appear to be sick, and the ratio between infections and fatalities (was) very high, particularly for older people and people with preexisting medical conditions" (Sovacool et al., 2020). Because of multiple infection waves, expanding death tolls, and unprecedented economic, health, and social effects, many academics and writers labeled the pandemic period as the "Era of Covid-19" (Bauchner and Fontanarosa, 2020; Flannery, 2020; Horton, 2020; Portnoy et al., 2020). However, the widespread and interconnected problems that existed during the period proved that the label was too narrow.

Coronavirus timeline

In December 2019, the pandemic began as a local outbreak, in Wuhan China, with dozens of cases. According to an article in *Science*, "there was a preponderance of early Covid-19 cases associated with Huanan Market," which sold live mammals susceptible to coronaviruses (Worobey, 2021). On December 30, 2019, the Wuhan Health Commission issued two warnings to local hospitals explaining the emergence of patients with unexplained pneumonia, several of

TABLE 1.1 Countries with the most confirmed cases at the end of 2021

Number	Country	Confirmed cases	Population	Confirmed cases as a percentage of population
1	United States	47,945,945	329,500,000	14.6%
2	India	34,587,822	1,380,000,000	2.5%
3	Brazil	22,080,906	212,600,000	10.4%
4	United Kingdom	10,189,063	67,220,000	15.2%
5	Russia	9,636,881	144,100,000	6.7%
6	Turkey	8,770,372	84,340,000	10.4%
7	France	7,394,153	67,390,000	11.0%
8	Iran	6,113,192	83,990,000	7.3%
9	Germany	5,836,813	83,240,000	7.0%
10	Argentina	5,326,448	45,380,000	11.7%
11	Spain	5,153,924	47,350,000	10.9%
12	Colombia	5,065,373	50,880,000	10.0%

Source: World Health Organization, Covid19.who.int, accessed December 1, 2021.

whom worked at the Huanan Market. The first public announcement came the next day, identifying 27 patients with coronaviruses (Worobey, 2021). On January 11, 2020, Chinese health officials recorded the first death from Covid-19 (Taylor, 2021). The victim was a regular customer at the Huanan Market. The report of the patient's death arrived before the Chinese New Year, in which hundreds of millions of people traveled across the country. Because of the holiday, the novel coronavirus spread throughout the province and beyond. On January 30, 2020, amid thousands of cases in China, the World Health Organization (WHO) declared the outbreak a public health emergency. In the meantime, many of those infected by the virus boarded planes, seeding outbreaks in the rest of the world. On March 11, 2020, the WHO declared the outbreak a global pandemic. At the end of 2020, 1 year into the crisis, the WHO recorded more than 65,000,000 global infections and 1,500,000 deaths. At the end of 2021, 2 years into the crisis, the WHO recorded more than 260,000,000 global infections and 5,200,000 deaths, with the United States, India, and Brazil topping the list (Table 1.1). Among the hardest-hit countries, the United Kingdom, United States, and Argentina had the highest number of confirmed cases as a percentage of the population.

A long tail of disruption

It is important to study the coronavirus pandemic as a shock that ravaged the institutions and processes of modern civilization. But it did not exist in isolation. The renowned historian Niall Ferguson (2021) argues that "a pandemic is not a single, discrete event. It invariably leads to other forms of disaster—economic, social, and political. There can be, and often are, cascades or chain reactions of disaster." Laura Robinson, Associate Professor of Sociology at Santa Clara University, and her coauthors (2021) provide a context:

> Unlike disasters that are more temporarily and spatially bounded, the pandemic...continued to expand across time and space . . . leaving an unusually broad range of second-order and third-order harms in its wake. . . . The pandemic . . . deepened existing inequalities and created new vulnerabilities related to social isolation, incarceration, involuntary exclusion from the labor market, diminished economic opportunity, life-and-death risk in the workplace, and a host of emergent digital, emotional, and economic divides. In tandem, many less advantaged individuals and groups . . . suffered disproportionate hardship related to the pandemic in the form of fear and anxiety, exposure to misinformation (false or inaccurate claims), and the effects of the politicization of the crisis. Many of these phenomena will have a long tail that we are only beginning to understand.

This book argues that the coronavirus pandemic intersected with crises in the economy, environment, and society. Because of the potential for future vulnerability, this reality is important: problems spread through multiple systems, potentially lingering for years. Even more, as the book explains, the outcomes of economic, health, and social crises alter the trajectory of society. As a result, the "Era of Cascading Crises" serves as a descriptive label for the period of time that includes but is not limited to Covid-19. The period encompasses six interconnected crises: coronavirus pandemic, economic collapse, climate catastrophe, racial injustice, domestic violence, and epistemic oppression. As Bruce Parrott (2020) of the Johns Hopkins School of Advanced International Studies argues in *Challenge*, the pandemic's "destabilizing challenges will be more diffuse and will vary in substance over time." With respect to the structure of the book, Chapters 2–7 address these crises, acknowledging their interconnection.

Coronavirus pandemic

Chapter 2 explains that the coronavirus pandemic and the international turmoil it triggered demonstrate the multi-dimensional nature of global upheaval. The disease Covid-19 ravaged human health and economic activity with ripple effects on every aspect of human life. Unprecedented lockdowns, travel restrictions, and periods of economic shutdown created recession and **contagion**—the spreading of harmful outcomes—across regions and continents. To respond to changing pandemic conditions, businesses adjusted their profit forecasts and methods of organization. Policymakers balanced the benefit of slowing the spread of the virus against the costs of recession, social instability, and human anxiety. Individuals altered their expectations, especially with employment, home life, and social interaction.

Economic collapse

Chapter 3 discusses the fact that, to slow the spread of the virus, economic shutdown interventions closed businesses, disrupted supply chains, and decreased production, leading to demand shocks, supply shocks, and global economic collapse (Bauer et al., 2020). Contraction ended years of economic growth, disproportionately impacting lower-wage and less-educated workers, minority members of society, and women. The interventions also impacted spending patterns. Although some industries (groceries, online retailers, and pharmacies) increased sales, others (bars, restaurants, aviation) experienced losses. In this context, an important question related to contraction is: how much of the economic decline resulted from government intervention or individuals voluntarily staying home to avoid infection? The University of Chicago economists Austan Goolsbee and Eric Syverson (2021) found that legal restrictions accounted for a small percentage of economic contraction: because of the growth in online commerce and the fear of infection, individual choice was more impactful than government mandates.

Climate catastrophe

Chapter 4 argues that climate change—long-term shifts in temperature and weather patterns—exists as the largest form of market failure in human history. In this context, the market's profit incentive leads to climate catastrophe. Climate problems result from fossil fuel combustion and greenhouse gas emissions: higher global temperatures (1°C increase since the first industrial revolution), rising sea levels, droughts in dry areas, flooding in wet areas, wildfires, and climate conflict. In an article in *Nature*, Timothy Lenton, director of the Global Systems Institute at the University of Exeter, and his coauthors (Lenton et al., 2019) argue that the idea of a **tipping point**—the point at which small changes in a system become large-scale discontinuities with irreversible outcomes—characterizes the climate problem. Examples include the melting of the West Antarctic ice sheet due to rising temperatures and the loss of Amazonian rainforest due to deforestation. To address the climate catastrophe, countries must reduce their greenhouse gas emissions, commit financial support to the developing world, and establish a framework of collective action.

Racial injustice

Chapter 5 discusses racial injustice that existed before, during, and after the coronavirus pandemic. During this period, the Black Lives Matter (BLM) movement served as the most prominent form of resistance. Although police brutality sparked BLM protests, higher levels of anxiety, tension, and economic instability during the coronavirus pandemic contributed to the problem. High-profile cases, including the George Floyd murder at the hands of a police officer, in Minneapolis, Minnesota, on May 25, 2020, existed as examples of a specific claim:

because of systematic racism, police in the United States disproportionately used lethal violence against African Americans. During the period of cascading crises, 25 million people around the world marched in support of BLM. Police brutality reveals the existence of systematic racism, when racism is embedded in customs, laws, and social behavior.

Domestic violence

Chapter 6 considers the rise of domestic violence during the coronavirus pandemic. Evidence demonstrates global patterns. Early in the pandemic, an article in *The Guardian* argued that "Women and children who live with domestic violence have no escape from their abusers during quarantine, and from Brazil to Germany, Italy to China, activists and survivors say they are already seeing an alarming rise in abuse" (Graham-Harrison et al., 2020). Lockdown measures led to a range of violations in domestic space. Multiple stresses from Covid-19, economic collapse, and social instability increased household anxiety. Victims of domestic abuse were stranded with abusers, unable to access support networks. In this environment, a pattern emerged. For women and children in volatile households, distorted power dynamics destabilized family environments. In the absence of legal and social forms of oversight, restrictions on movement eliminated avenues of escape: "lockdown measures may . . . grant people who abuse greater freedom to act without scrutiny or consequence" (Bradley-Jones and Isham, 2020).

Epistemic crisis

Chapter 7 explains that an epistemic crisis—a crisis of knowledge—interacts with pandemics. Individuals may use social media platforms to propagate misinformation, believing "different versions of reality based on the digital communities in which they congregate" (Manjoo, 2022). Examples include the false beliefs that all scientists exaggerate pandemic risks, vaccines are ineffective, and individuals should not alter their behavior to reduce the risk of infection. Misinformation destabilizes democratic processes, exacerbates divisions, and erodes trust in public institutions (Benkler et al., 2018). During a pandemic, when society needs collective action to both stop the spread of disease and initiate a process of recovery, an epistemic crisis prolongs negative health outcomes by propagating false beliefs. The convergence of artificial intelligence, big data, and social media—beyond the control of governing authorities—creates digital echo chambers that reinforce pre-existing biases. These problems weaken social cohesion and trustworthiness, reducing the ability of society to recover from disruption.

Thesis and themes

The book's thesis is that the coronavirus pandemic existed as one crisis in an interconnected series of cascading crises, including economic collapse, climate change, racial injustice, domestic violence, and epistemic oppression. The book

TABLE 1.2 Themes of the book

Topic	Theme
Characteristics of crises	Crises may be linear or cyclical, momentary or enduring, singular or recurrent, and systematic or episodic.
Networks	The larger the number of network connections, the greater the potential for chain reactions.
Unintended consequences	Unintended policy consequences may lead to economic and social instability.
Intersectionality	Intersectional categorizations such as race, class, and gender create interdependent forms of disadvantage.
Poverty, discrimination, and inequality	During periods of instability, vulnerable members of society bear a disproportionate burden.
Systems approach	A systems approach addresses statistical patterns in nondiscriminatory ways.
Discord	Declining health, economic contraction, climate chaos, and social instability contribute to an age of discord.

Source: Author.

aims to equip students with the conceptual clarity to understand global crises, assess policy interventions, and analyze social responses. As Ferguson (2021) argues, a pandemic includes three elements, the pathogen, process of contagion, and systems under attack, and so "We cannot understand the scale of the contagion by studying only the virus itself, because the virus will infect only as many people as social networks allow it to." Using this framework, *Understanding Global Crises* explores several themes (Table 1.2).

Characteristics of crises

Crises come in different forms. They exist in global, national, regional, and local dimensions. They are uncertain with respect to their timing. They impact economic, health, political, and social systems. Colin Hay (1999) argues that crises may be linear or cyclical, momentary or enduring, pathological or regenerative, singular or recurrent, and systematic or episodic. Depending on the country or region, crises may exist as rare or repetitive events. But when they occur infrequently, they are difficult to imagine. The complexity of modern societies makes them susceptible to crises, including pandemics, wars, and upheavals. In this context, Bruce Parrott (2020) makes two points. First, the retrospective identification of crises depends on analysts documenting previous events, especially with respect to chronology, spatial boundaries, and intersection. The Great Recession began in the United States, occurred in 2008–2009, and spread in financial networks. The coronavirus pandemic began in China, occurred in 2020–2022, and spread in global transmission networks. Second, crises share characteristics (Table 1.3). Because of this interconnection, government responses may entail diffuse responsibility, intellectual inadequacy, or ideological incompetency. Even after societies emerge from a period of disruption, economic, political, and social instability may linger.

TABLE 1.3 Characteristics of crises

Characteristic	Explanation
Consequences	Recovery plans lead to painful consequences.
Expectations	Disruption emerges with unexpected severity.
Order	Economic and social outcomes alter the existing order.
Policy	Policy options may not offer ready-made solutions.
Threats	Disruption alters economies, political systems, and societies.
Values	Instability provokes debate about which values to protect.

Source: Parrott (2020).

Creative, enduring, and novel possibilities

Crises also generate creative, enduring, and novel possibilities. They provide opportunities for **innovation** when necessity fosters creation. Innovation, a new idea, process, or product, stems from the desire to create markets, methods, and output. Strategic and serendipitous behavior produces new outcomes, including remote work, teleconferencing, and delivery on demand. A time of hardship creates an intermediate state of affairs between periods of relative stability.

Transformative change

Transformative change involves alterations to patterns of activity. It affects production, consumption, resource allocation, and public policy. Transition is the period of time between the introduction of new ideas, methods, and products and their implementation. An external shock such as a pandemic accelerates the transition. Before the coronavirus pandemic, for example, remote work was an option provided by progressive companies to recruit talented employees: only 7 percent of civilian workers in the United States had access to flexible workplaces (Desilver, 2020). But digital interaction, file sharing, and teleconferencing enhanced market connectivity, increasing remote work options in multiple industries (Yang et al., 2021). Before the end of the pandemic, the category of remote workers expanded to more than 42 percent of employed workers in the U.S. economy, according to the Bureau of Labor Statistics (Kessler, 2022).

Potential for decline

As Chapter 8 explains, while crises may lead to creative, enduring, and regenerative effects and the potential for transformative change, they may also lead to decline. Modern, global, and interconnected societies, even powerful ones, are vulnerable. Disruption of the existing order has been a recurrent theme throughout history, and major disruptions accelerate the trend. Depending on the severity of the crisis and effectiveness of the economic, political, and social response, decline may manifest in the form of less organization and coordination of individuals and groups, rising inequality, a decrease in accurate information between a center and periphery, less investment in the elements that define civilization, and many other factors.

A moment of truth and revelation

Meeting unprecedented demands requires comprehensive systems of risk assessment, flexible public sector policies, sophisticated forecasting and planning, and the establishment of priorities. Crises require broad interpretations of the common good, presenting an enormous challenge for democracies (Parrott, 2020). In effect, a pandemic is "a moment of truth, of revelation, exposing some (countries) as fragile, others as resilient, and others as antifragile—able not just to withstand disaster but to be strengthened by it" (Ferguson, 2021).

Moving forward

Successful crisis management depends on accurate and timely diagnoses, the availability of appropriate expertise, implementation of collective action, material resources, and competent leadership. If society cannot convey important information, conflicts may erupt over the origin of crises and who bears their burden. Crises, therefore, ignite debates over problem characteristics, potential solutions, and future pathways. Moving forward, society does not know when a future pandemic will strike. But it is important to learn from past experience, build resilient institutions, and avoid descension into confusion and instability.

Questions

This analysis of truth and revelation raises important questions. Why do some societies manage crises better than others? Why do some countries crumble, some maintain the status quo, and others emerge stronger? While maintaining capacity and security, is it possible to minimize economic and social instability? During crises, what motivates the choice of individualism or collective action? Do attitudes and beliefs on poverty, discrimination, and inequality matter? These are the central questions posed by *Understanding Global Crises*.

Networks

Disruptions normally exhibit limited geographic reach, impacting a community or a region. In the twentieth century, the world wars were exceptions. "What matters is, first, whether or not a disaster strikes a densely populated part of the earth and, second, if the death and destruction in and around the epicenter have repercussions further afield" (Ferguson, 2021). That is, large-scale crises result from contagion, a process that pushes an initial shock through networks, including human transmission networks, social networks, and supply chain networks. In form, networks consist of nodes (points in which connections intersect or branch) and relationships between nodes. New connections increase network flow, enhancing the likelihood that an initial disturbance leads to cascading effects. During a pandemic, societies fight the spread of disease by minimizing the impact of an infectious pathogen. But networks—especially human transmission networks—are as important in determining pandemic outcomes as the infectiousness of the pathogen. The reason is that illness and fatality are not inherent to a new virus. Illness and fatality are as much a function of network connections as epidemiological realities.

Network characteristics

Networks arise for different reasons, including international trade, market connection, and social interaction. Areas of differentiation include the degree of self-organization, structure, and speed of transmission. Power laws, propagation of change, and systematic risk characterize networks.

Power laws

Power laws establish functional relationships between quantities when a relative change in one quantity leads to a proportional change in the other. Many examples of power laws exist, including income disparities, stock market returns, and frequencies of words in languages. Power laws reveal correlations between different variables. Among connected parts, small changes may lead to cascading effects, such as knocking down the first domino in a row or bowling pin in a set (Arthur, 2021).

Propagation of change

Topologies and densities impact network flow. In a sparsely connected network, disturbances dissipate from a lack of onward transmission. But in a densely connected network, disturbances advance. The network may experience a phase change when transition occurs from a state with certain parameters to a different configuration. In this case, network flow may progress from a few to many consequences (Arthur, 2021).

Systematic risk

Risk means exposure to danger, harm, or loss. Different categories of risk exist, including low, medium, and high. Standard methods of risk management categorize low-risk disruptions, such as wind damage, with little catastrophic potential or geographic dispersion. Standard methods of risk management struggle with intermediate-risk disruptions, such as regional disease outbreaks. Standard

TABLE 1.4 Risk criteria

Number	Risk criteria	Explanation
1	Damage effects	Outcomes measured in quantifiable units
2	Delay effects	Time between the triggering event and onset of damages
3	Incertitude	Degree of uncertainty or lack of probability
4	Inequity	Who bears the burden
5	Mobilization	Potential to address disruption in a collective manner
6	Persistency	Length of disruption
7	Probability	Likelihood of disruption as discrete or continuous loss
8	Reversibility	Potential to restore the existing order
9	Source	Origin of disruption
10	Violation	Generation of additional problems
11	Ubiquity	Geographic dispersion of damage

Source: Renn and Klinke (2004).

methods of risk management struggle the most with high-risk disruptions, irreversible damage effects, continuous loss, and wide geographic dispersion, including global pandemics.

Networks experience systematic risk. Systematic risk combines disruptive events and socioeconomic factors, focusing on the interdependencies and connections between disruptive events, human behavior, and institutions. It merges crisis identification, risk assessment, and risk management. For example, if a group of nodes, such as banks in a financial market, are independent and unconnected, a disturbance in one part of the network will not spread in a process of contagion. If, however, the nodes operate in an interconnected network of transmission, contagion may occur. In the latter case, risk assessment would determine that a crisis starting in one part of a network could spread throughout, threatening overall collapse, such as the banking system during the Great Recession of 2008 or the healthcare system during the coronavirus pandemic of 2020 (Arthur, 2021).

According to Ortwin Renn and Andreas Klinke (2004), risk criteria establish a framework of analysis (Table 1.4). They demonstrate why a shock disrupts certain networks, creates uneven outcomes, and persists in specific segments of the population. But the violation criterion requires explanation. It consists of four sub-factors: injustices associated with costs, benefits, and social status; psychological stress and discomfort from damage effects; potential for social conflict; and spillover effects that impact other areas in a process of contagion (Renn and Klinke, 2004).

Network transmission

The degree of severity of a disease outbreak depends on the interaction between pathogens, carriers, and human transmission networks. Network structure is as important as the pathogen. When an infectious pathogen goes viral, the key is the connection between nodes (contagious individuals) and the rapidity with which they infect others. The process depends on the susceptibility of individuals and whether interventions break transmission chains. A few highly connected nodes, especially superspreaders—those who transmit an infectious disease to many others—may cause a local outbreak to become a contagious event.

Degree of connection

A complex, global pandemic, different from a local outbreak, requires a critical mass of infections and a high degree of human connection. While an initial disruption influences the outcome, network structure facilitates transmission. In reality, for every disease outbreak that becomes a national epidemic or global pandemic, countless others diminish into obscurity. In the latter cases, disturbances impact isolated nodes in dispersed networks, failing to generate larger outcomes.

Increasing vulnerability

Evolving processes impact humanity's vulnerability to infectious diseases. First, urbanization, an increase in demographic and economic concentration, leads to greater population density and potential for contagion. Second, industrial agriculture creates an economic structure in which a large number of people do not have to work in food production, diversifying the economy and creating new systems of interconnection. Third, **globalization**, the world's growing interconnections, creates global networks of exchange. Together, these factors demonstrate that the more society increases population density, enhances systems of interconnection, and strengthens human transmission networks, the more vulnerable it is to global pandemics.

Unintended consequences

In 2020, when the novel coronavirus was spreading around the world, infections and deaths were rising, and hospitals were reaching capacity, a vaccine was not yet available. Most countries implemented lockdowns and targeted quarantines, canceled flights, closed borders, shuttered non-essential workplaces, encouraged remote work, prevented large gatherings, and required social distancing. Individuals sheltered with a pod of personal connections. But problems escalated, including the restriction of movement and confinement of the sick with the healthy. Over time, the most deleterious consequences related to economic and social instability. The key to understanding this reality is the concept of **opportunity cost**, the value of the best foregone alternative.

Opportunity cost

The transmission of a novel coronavirus is highly complex, due to the nonlinear nature of infections. However, when lockdown policies restrict interaction, they slow the spread of disease (Yin et al., 2021; Flaxman et al., 2020; Hsiang et al., 2020). The problem is the opportunity cost of this choice. Ananish Chaudhuri (2022), Professor of Experimental Economics at the University of Auckland, argues that, when countries divert scarce resources to enforce lockdowns, opportunity cost exists. In this situation, society may not allocate a sufficient level of resources to help those who suffer from anxiety, loneliness, and unemployment. First, a study by Harvard University found that, during the first year of the pandemic, one-third of Americans described themselves as "seriously lonely," up from one-fifth before the pandemic (Weissboard et al., 2021). Second, during the first year of the pandemic, the number of alcohol-related deaths in the United States, including from accidents and liver disease, increased by 25 percent (Rabin, 2022). Third, during the pandemic, people in the United States died of drug overdoses in record numbers (Rabin, 2021). Fourth, lockdown policies led to an increase in domestic violence in countries around the world (Aguero, 2021). These and other examples demonstrated the external cost of solitude. Unable to leave, many individuals experienced their environments as lonely and dangerous places.

The reality of unintended consequences

Policy evaluation must include both costs and benefits. What are the objectives of lockdowns? When they occur, what does society sacrifice? The challenge, according to Chaudhuri (2022), is that governing authorities tend to focus on policy benefits, including lives saved or infections avoided, but not the costs that are more "diffuse and happen in a more dispersed manner." In Chaudhuri's framework, when society allocates resources to achieve specific goals, such as increasing hospital capacity or decreasing the spread of a virus, it diverts resources from other areas, including the economy and society. The result is a series of unintended consequences. During the coronavirus pandemic, unintended consequences manifested in many forms, including a rise in racial injustice, domestic violence, and epistemic oppression.

Shane Sanders (2020), Professor of Sports Analytics at Syracuse University, argues that policies often have unintended consequences. In his book, *The Economic Reason*, Sanders emphasizes that the field of economics addresses many well-understood scenarios: seatbelt mandates lower the effective cost of reckless driving and increase the incidence of accidents; decreasing the price of abortion, legalization causes a future decrease in crime rates by limiting the number of at-risk children being born; in basketball, an increase in distance of the three-point line causes the percentage of both three-pointers and two-pointers to decline.

Many other examples exist. The point is that it is important to consider the universal nature of unintended consequences: policies may "cause targeted problems to persist and untargeted problems to diminish without apparent explanation" (Sanders, 2020). But the opposite may also hold true. Policies, especially lockdowns, may solve targeted problems (reducing the spread of a virus) but cause others to fester (racial injustice, domestic violence, and epistemic oppression).

Intersectionality

Kimberle Crenshaw (2013), a pioneering scholar on civil rights and Professor at Columbia Law School, coined the term **intersectionality** to "denote the various ways in which race and gender interact to shape the multiple dimensions" of the lives of Black women. But she expanded the concept, arguing that the theory is not exclusive. Rather, it applies to all members of society that experience multiple forms of oppression. Because of discrimination and inequality, society has traditionally placed economic and social roadblocks in the paths of women, people of color, individuals with non-heterosexual forms of identity, and other oppressed groups. Crenshaw's (2013) theory implies that, if individuals experience multiple layers of oppression, their places in the existing order "cannot be captured wholly" by evaluating their experiences separately. Their places are a function of their multiple identities.

Crenshaw's example

During a field study of battered women's shelters in minority neighborhoods in Los Angeles, Crenshaw (2013) observes an intersectional problem: the physical assault on women in cases of domestic violence that exist as the "most immediate manifestation of the subordination they experience." Many of the victims, however, also experience unemployment and poverty. In addition, many suffer from abusive relationships, a lack of job skills, and immigrant status. In the shelters, discriminatory practices compound problems of race, class, gender, and socioeconomic status. It is, therefore, important for the shelters to confront "multilayered and routinized forms of domination that often converge" in the lives of the battered women (Crenshaw, 2013).

Applications of intersectionality

The theory of intersectionality engages the dominant assumption that race, class, and gender exist as exclusive or separate categories. Crenshaw (2013) argues that these factors interact. In addition, they create synergistic effects, placing individuals suffering from multiple forms of oppression (for example, immigrant women of color) at a disadvantage relative to those with one form of oppression (white women) or no forms of oppression (white men). According to Crenshaw, the theory also includes age and sexual orientation, so it establishes a comprehensive framework.

In *Understanding Global Crises*, analysis of the intersection of multiple forms of identity provides a method to evaluate the outcomes of economic, health, and social instability. During the coronavirus pandemic, labor market problems and support mechanisms affected individuals in different ways. For example, the economist Thomas Crossley and his coauthors (2021) argued that, during the first infection wave, "labour market impacts were most negative for individuals from minority ethnic groups and those in the lowest quintile of average pre-Covid-19 income."

An intersectional framework raises important questions. During the pandemic, if no one had natural immunity to the novel coronavirus, why did it disproportionately impact people of color and members of lower socioeconomic classes? Why did the shutdown interval impact women more than men? During the period of social instability, why did certain members of the population suffer more from racial injustice and domestic violence? To address these questions, the book uses an intersectional framework.

Poverty, discrimination, and inequality

An important theme in *Understanding Global Crises* is that the costs of a pandemic—in terms of rising unemployment, morbidity, mortality, and social instability—are borne disproportionately by the most vulnerable members of society. When pandemics strike, they lead to asymmetric costs. First, members of low-income households are more likely to experience chronic diseases, receive less healthcare services, and suffer from harmful health outcomes. They are ill-suited to address multifaceted threats to their personal well-being. During the pandemic, excess deaths were due to Covid-19 but also pre-existing conditions such as diabetes and heart disease. Second, discrimination—unjust or prejudicial behavior and treatment—includes racism and sexism and exists in institutions such as the economy, healthcare, housing, and law enforcement. During periods of social instability, the problems of discrimination and prejudice lead to inferior outcomes. Third, those with low-income status struggle to withstand a shock to the economy. Together, the factors of poverty, discrimination, and inequality contribute to unequal effects.

Costs to public health

In a policy brief, Luiza Nassif-Pires, a research fellow in the Gender Equality and the Economy program at the Levy Institute of Bard College, and her coauthors (Nassif-Pires et al., 2020) argue that education, ethnicity, income, occupation, and race correlate with the incidence and severity of disease, including Covid-19. For vulnerable members of society, this link may be catastrophic. To establish a relationship between poverty and the clinical risk of disease, Nassif-Pires et al. (2020) develop a health risk index. The index demonstrates that residents of poorer neighborhoods experience higher rates of asthma, cancer, diabetes, and kidney disease. These problems exacerbate pandemic outcomes: "In neighbor-hoods where the share of the population living below the poverty line is 45 per-cent or greater, the risk factor index is above the national average" (Nassif-Pires et al., 2020). In addition to poverty, increased health risk stems from inadequate access to healthcare, differences in living conditions, and an absence of sick-leave policies.

Costs to economic well-being

During crises, the most economically vulnerable members of society struggle to maintain their levels of subsistence living. A first area of concern relates to necessities: lower-income households may not have sufficient food and clothing. During lockdown, they may not be able to minimize their grocery store trips, in-creasing exposure to the pathogen. In contrast, those with greater economic means work from home, avoid public spaces, and have the resources to insulate themselves from the onslaught of disease. A second area of concern relates to services. When schools close during the shutdown interval of the pandemic, childcare moves inside the home. Households compensate by increasing the number of hours they spend supervising the education of children, which complicates work responsibilities. A final area relates to gender. Women assume a disproportionate share of house-hold responsibilities. Women allocate more time than men for childcare, cooking, cleaning, and shopping. Even more, unemployed men often resist domestic chores, placing greater burden on working women. When households struggle, women assume greater responsibility (Nassif-Pires et al., 2020).

Case study 1.1 Poverty, intersectionality, and unequal outcomes

During the coronavirus pandemic, interrelated inequalities in the deter-minants of health—access to care, discrimination, education, housing, in-come, occupation, and wealth—led to disproportionate effects for the most vulnerable members of society. These disadvantages placed them at the greatest risk for infection and unemployment. While inequalities in health and economic outcomes were the byproduct of historical systems of disad-vantage, the coronavirus pandemic exacerbated these problems.

In the United States, when the vaccine in late 2020 and early 2021 inoculated frontline workers and people in nursing homes, a pattern emerged. Although low-income, marginalized communities of color were hit hardest by the pandemic, individuals from wealthier and mostly white neighborhoods in Dallas, Miami-Dade County, New York City, Washington D.C., and other cities were reserving vaccine appointments in poorer neighborhoods. During this period, the demand for vaccines exceeded supply. People from all socioeconomic backgrounds were anxious to receive inoculations. But wealthier individuals flooded vaccination centers, creating obstacles for individuals in underserved neighborhoods: "registration phone lines and websites . . . can take hours to navigate, and lack of transportation or time off from jobs (impact the ability) to get to appointments" (Goodnough and Hoffman, 2021).

In New York City, with a population 24 percent Black, 29 percent Latinx, and 43 percent white, inoculation data revealed that, for the recipients of the first 300,000 doses, 11 percent were Black, 15 percent were Latinx, and 48 percent were white (Fitzsimmons, 2021). While the vaccines over-represented the white population, they underrepresented Black and Latinx communities.

This trend prevented a sufficient supply of vaccinations from flowing to those with pre-existing medical conditions and less access to healthcare. Although many cities, including New York, addressed this problem with outreach campaigns, vaccination centers in housing complexes, phone lines to complement online registration, and prioritization of appointments for people in neighborhoods with the highest infection rates, unequal access to vaccinations in the initial rollout served as an example of an inequality of outcomes.

County data from Pacoima, a Latinx neighborhood in Los Angeles, revealed a rate of Covid-19 cases five times the rate in whiter and richer Santa Monica, 25 miles to the south. In Los Angeles, essential employees making deliveries, working in warehouses, and serving in food production were more likely to be Latinx, become infected at work, live in overcrowded housing, and facilitate the spread of the virus (Cowan and Bloch, 2021).

In Chicago, a city with a 32.4 percent African American population, 61 of the first 86 recorded deaths (70 percent) were African Americans. Of the first 3,000 Chicagoans who died from Covid-19, 40 percent were African Americans. Tabulated in another way, during the first year of the pandemic, 16.1 deaths per 10,000 residents occurred with African Americans. But, for white Chicagoans, 9.2 deaths per 10,000 residents occurred (Zamudio, 2020).

During the coronavirus pandemic, New York, Los Angeles, Chicago, and many other cities around the world struggled to provide care for households with low socioeconomic status. Governing authorities struggled to allocate scarce resources to overstretched and overwhelmed healthcare systems. At the same time, resources were slow to trickle down to impoverished neighborhoods. Injustices of the past, in other words, repeated during the coronavirus pandemic. Both the marginalization of poor communities and structural racism perpetuated inferior health outcomes (Ivers and Walton, 2020).

Systems approach

To analyze the cascading crises in a multi-disciplinary framework, the book uses a systems approach. This approach addresses statistical patterns in nondiscriminatory ways. It also acknowledges the important role of socioeconomic patterns. As the philosophers Nadya Vasilyeva and Saray Ayala-Lopez (2019) argue, a systems approach explains "why certain social groups are overrepresented or underrepresented in specific domains, without assuming inevitability and normativity of the existing patterns."

Attributes of a systems approach

First, a systems approach identifies the characteristics of social systems. The existing order, for example, may restrict the ability of vulnerable members of the population to increase their living standards. Second, the approach identifies structural constraints, the factors that complicate rational decision-making. These constraints—including poverty, discrimination, and inequality—act on the characteristics of the system to shape the distribution of outcomes. As Chapter 2 explains, during a pandemic, individuals with higher levels of education and income experience better health. These factors decrease the risk of exposure to an infectious pathogen. Third, a systems approach addresses social constructs such as race, class, and gender, rather than inherent aspects of their categorical essence. Chapter 7 argues that this approach applies to epistemic injustice when individuals experience inaccuracies or falsehoods as knowers or transmitters of knowledge, due to discrimination, prejudice, or stereotype.

> When attributes of social groups are seen as fixed and derived from an internal core, people are likely to rely on stereotypes to navigate the social reality. As long as structural thinking takes us away from the sort of cognitive processes that lead to problematic judgments of other people, developing structural thinking has something to offer to the fight against epistemic injustice.
>
> *(Vasilyeva and Ayala-Lopez, 2019)*

Structures

Structures such as the economy exist as **complex systems** with competition, dependencies, and evolving connections. Because asymmetric organization and interacting components characterize complex systems, they operate between conditions of order and disorder. Order means the arrangement of people or things in relation to each other in patterns or sequences. Disorder is the absence of this arrangement. The existence of a specific characterization means a system may operate in equilibrium, experience a shock, reach a critical stage, decline, and then collapse. A pandemic may trigger transition from stability to chaos. But the system may also regenerate.

Complex adaptive systems

The economy, Internet, and other forms of organization possess features of complex adaptive systems. The economist W. Brian Arthur (2021) of the Santa Fe Institute, a research center for complex systems science, argues that dispersed interaction between multiple agents characterizes complex adaptive systems. These systems innovate, evolve, and possess specific features:

- A disturbance in one part of the system may create cascading effects throughout.
- A nondeterministic possibility may create nonlinear relationships.
- A range of potential shocks may prevent measurement of the scale of disruption.

With interconnected systems, these features mean that a shock such as disease outbreak may lead to disproportionate and cascading effects. However, even in complex adaptive systems, disruptions and crises may initiate or perpetuate an age of discord.

Age of discord

Peter Turchin (2016), one of the world's experts on the functioning and dynamics of historical societies, argues that deep and structural problems contribute to crises. Because historical analysis is a function of the conditions of the period, it is challenging to establish historical patterns. However, Turchin contends that, with sufficient data and historical oversight, it is possible to make observations. Specifically, Turchin argues that, in the process of economic development, progress and material success may give way to social maladies, including income gaps and ineffective government. In Turchin's framework, progress and material success create a bloated class of elites, rising income and wealth inequality, and challenging living conditions. Over-extension by the public sector creates an inability to cover its financial position, leading to ineffective governance. These problems create the potential for social discord.

Expansion and decline

As economic development occurs, pressure exists for social cohesion (Turchin, 2003). But when a country establishes an advanced level of civilization—with luxuries, socioeconomic classes, and prejudices—both cooperation and social cohesion decline. In a process of change, government, population, social structure, and sociopolitical stability are important. Initially, continuity and peace lead to prosperity. But, when rising income and wealth inequality benefit the rich at the expense of everyone else, economic growth creates instability. Those who suffer from economic and social maladies protest the existing order. An overextended government and a bloated elite sector do not maintain the fiscal balance necessary to address ongoing crises. Eventually, ineffective government, social instability, and income inequality lead to a period of decline.

The role of information

In this framework, noncyclical forces, including pandemics, inject elements of disruption. But Jaeweon Shin of Rice University and his coauthors (2020) propose an important addition: a society's process of development is governed first by the growth of civil government and economic systems and then information processing. Eventually, a scale threshold of development emerges, beyond which growth in information technology serves as the most important factor. Shin et al. (2020) ask:

> Could some of the frequent collapses seen in societies be due to a polity's never developing sufficient information-processing capacities, so that it stumbles or even collapses through poor performance due to lack of external connectivity, internal coherence, or inability to compete with polities whose superior information-processing abilities have enabled more growth in size?

In a period of political division, economic inequality, climate instability, and social discord, after a shock disrupts the existing order, an inadequate informational response may initiate a gradual process of decline.

Application

Analysis of the contemporary global environment reveals interconnected networks, positive and negative global flows, and human civilization vulnerable to the spread of infectious disease. During a pandemic, individuals with the economic means work from home or decamp to the countryside. But the working poor maintain their frontline positions, experiencing the worst of the pandemic. Governments struggle to implement an effective system of testing, contact tracing, and targeted quarantines. The proliferation of misinformation weakens the potential for collective response. Countries with ineffective leaders and institutions struggle to care for the poor and dispossessed. The result is a decline of the existing order, reflected in division, misinformation, and stratification.

The need for collective action and cooperation

Chapter 9 argues that collective action and cooperation exist as effective methods to address global pandemics, economic collapse, climate instability, and social discord. Turchin (2016) agrees: "Societal breakdown and ensuing waves of violence can be avoided by collective, cooperative action." But societies may lack the social capital to act in a collective and cooperative manner. Social capital refers to the networks of relationships that enable society to function. In the presence of inequality, misinformation, and political division, societies struggle.

During the coronavirus pandemic, for example, intervention measures such as mask wearing, social distancing, and vaccinating served as crucial policies to slow the spread of disease. But the adoption of these measures varied across and within countries, depending on the preference for individualism, competence of leadership, presence of misinformation, and balance between the tradeoffs of policy design.

In an era of cascading crises, the interconnected nature of these factors determines outcomes: "A human society is a dynamical system, and its economic, social, and political subsystems do not operate in isolation" (Turchin, 2016). Modern-day societies possess the theories, data, and policies to function effectively; however, in the presence of the interconnected forces of individualism, inequality, and instability, societies may cycle between periods of progress and discord.

Summary

The coronavirus pandemic existed as a global shock, traveling through transmission networks, ravaging human health, increasing morbidity and mortality, leading to lockdowns and economic shutdowns, and creating social instability. But the pandemic, which led to hundreds of millions of global infections and millions of deaths, existed as one of the several cascading crises, including economic collapse, climate catastrophe, racial injustice, domestic violence, and epistemic oppression. In general, crises may be linear or cyclical, momentary or enduring, pathological or regenerative, singular or recurrent, and systematic or episodic. However, with a pandemic, human transmission networks are as important in determining outcomes as the infectiousness of the pathogen. The reality of unintended consequences means that opportunity cost exists: lockdown measures increase social instability. An intersectional framework acknowledges that multiple layers of oppression interact with economic, health, and social instability. Because of poverty, discrimination, and inequality, the most vulnerable members of society experience a disproportionate burden. A structural approach demonstrates that society may misrepresent certain groups in specific domains but does not assume inevitability of existing patterns. In this context, human societies and their dynamic systems do not operate in isolation. As a result, periods of growth and stability may transition through waves of discord and decline.

Chapter takeaways

LO1   Because the era of cascading crises involves the pandemic, economic collapse, climate catastrophe, racial injustice, domestic violence, and epistemic oppression, it exists as a disruptive period.

LO2 Crises may exist in global or national contexts, exhibit unpredictable patterns, and impact economic, health, political, and social systems.

LO3 When networks exhibit a high degree of connectivity, they are as important in contributing to a pandemic as the infectiousness of a pathogen.

LO4 The unintended consequences of lockdown interventions include, but are not limited to, a rise in loneliness, overdoses, and domestic violence.

LO5 For individuals, interdependent forms of disadvantage relating to race, class, gender, and socioeconomic status create disproportionate forms of hardship.

LO6 Poverty, discrimination, and inequality complicate economic, health, and social outcomes.

LO7 A systems approach reveals socioeconomic patterns.

LO8 When political division, misinformation, and uncertainty persist, a society may experience a period of decline.

Key terms

Complex systems	Intersectionality
Contagion	Networks
Crises	Opportunity cost
Globalization	Shock
Innovation	Tipping point

Questions

1 Do large-scale and interconnected problems characterize the current era? In your answer, provide examples.

2 List and explain the characteristics of crises. How do the characteristics relate to pandemics?

3 In the context of a pandemic, why are human transmission networks as important as the infectiousness of a virus?

4 During the coronavirus pandemic, explain the opportunity cost of lockdown policies. When lockdowns occur, what are the consequences? Should policymakers consider the potential for unintended consequences?

5 How does an intersectional framework provide insight on the economic, health, and social problems that emerge during a pandemic?

6 During a pandemic, how do poverty, discrimination, and inequality contribute to unequal outcomes?

7 Explain how complex systems operate between conditions of order and disorder. With respect to a pandemic, how does the concept of complex adaptive systems alter the framework of analysis?

8 For a society, what are the characteristics of structural decline? How does a lack of collective behavior contribute to the process?

References

Aguero, Jorge. 2021. "Covid-19 and the rise of intimate partner violence." *World Development*, 137: 105217.

Arthur, W. Brian. 2021. "Foundations of complexity economics." *Nature Reviews Physics*, 3: 136–145.

Bauchner, Howard and Fontanarosa, Phil. 2020. "Thinking of risk in the era of Covid-19," *JAMA*, 324(2): 151–153.

Bauer, Lauren, Broady, Kristen, Edelberg, Wendy, and O'Donnell, Jimmy. 2020. "Ten facts about Covid-19 and the U.S. economy." *Brookings*, September 17.

Benkler, Yochai, Faris, Robert, and Roberts, Hal. 2018. *Network Propaganda: Manipulation, Disinformation, and Radicalization in American Politics*. Oxford: Oxford University Press.

Bradley-Jones, Caroline and Isham, Louise. 2020. "The pandemic paradox: The consequences of Covid-19 on domestic violence." *Journal of Clinical Nursing*, 29(13–14): 2047–2049.

Chaudhuri, Ananish. 2022. *Nudged into Lockdown? Behavioral Economics, Uncertainty, and Covid-19*. Northampton, MA: Edward Elgar Publishing.

Cowan, Jill and Bloch, Matthew. 2021. "In Los Angeles, the Virus Is Pummeling Those Who Can Least Afford to Fall Ill." *The New York Times*, January 29.

Crenshaw, Kimberle. 2013. "Intersectionality and identify politics: learning from violence against women of color." In Kolmar and Bartkowski (Eds.), *Feminist Theory: A Reader*. New York: McGraw-Hill.

Crossley, Thomas, Fisher, Paul, and Low, Hamish. 2021. "The heterogeneous and regressive consequences of Covid-19: Evidence from high-quality panel data." *Journal of Public Economics*, 193: 104334.

Desilver, Drew. 2020. "Before the Coronavirus, Telework Was an Optional Benefit, Mostly for the Affluent Few." *Pew Research Center*, March 20.

Ferguson, Niall. 2021. *Doom: The Politics of Catastrophe*. New York: Penguin Press.

Flannery, Tim. 2020. *The Climate Cure: Solving the Climate Emergency in the Era of Covid-19*, Melbourne: Text Publishing Company.

Flaxman, Seth, Mishra, Swapnil, Gandy, Axel...Bhatt, Samir. 2020. "Estimating the effects of non-pharmaceutical interventions on Covid-19 in Europe." *Nature*, 584: 257–261.

Fitzsimmons, Emma. 2021. "Black and Latino New Yorkers Trail White Residents in Vaccine Rollout." *The New York Times*, January 31.

Goodnough, Abby and Hoffman, Jan. 2021. "The Wealthy are Getting More Vaccinations, Even in Poorer Neighborhoods." *The New York Times*, March 4.

Goolsbee, Austan and Syverson, Eric. 2021. "Fear, lockdown and diversion: Comparing drivers of pandemic economic decline 2020." *Journal of Public Economics*, 193: 104311.

Graham-Harrison, Emma, Giuffrida, Angela, Smith, Helena and Ford, Liz. 2020. "Lockdowns around the world bring rise in domestic violence." *The Guardian*, March 28.

Hay, Colin. 1999. "Crisis and the structural transformation of the state: interrogating the process of change." *British Journal of Politics and International Relations*, 1(3): 317–344.

Horton, Richard. 2020. "Offline: science and politics in the era of Covid-19." *The Lancet*, 396, 10259.

Hsiang, Solomon, Allen, Daniel, Annan-Phan, Sebastien...Wu, Tiffany. 2020. "The effect of large-scale anti-contagion policies on the Covid-19 pandemic." *Nature*, 584: 262–267.

Ivers, Louise and Walton, David. 2020. "Covid-19: Global Health Equity in Pandemic Response." *American Journal of Tropical Medicine and Hygiene*, 102(6): 1149–1150.

Kessler, Sarah. 2022. "No Prying. No Belly Pats. Just Work." *The New York Times*, March 6.

Lenton, Timothy, Rockstrom, Johan, Gaffney, Owen, Rahmstorf, Stefan, Richardson, Katherine, Steffen, Will, and Schellnhuber, Hans. 2019. "Climate tipping points—too risky to bet against." *Nature*, 575: 592–596.

Manjoo, Farhad. 2022. "We Might Be in a Simulation. How Much Should That Worry Us?" *The New York Times*, January 26.

Nassif-Pires, Luiza, Xavier, Laura, Masterson, Thomas, Nikiforos, Michalis, and Rios-Avila, Fernando. 2020. *Pandemic of Inequality*. Annondale-On-Hudson: Levy Institute of Bard College.

Parrott, Bruce. 2020. "The American mega-crisis: Covid-19 and beyond." *Challenge*, 63(5): 245–263.

Portnoy, Jay, Waller, Morgan, and Elliott, Tania. 2020. "Telemedicine in the era of Covid-19." *The Journal of Allergy and Clinical Immunology*, 8(5): 1489–1491.

Rabin, Roni. 2022. "Alcohol-Related Deaths Spiked During the Pandemic, a Study Shows." *The New York Times*, March 22.

Rabin, Roni. 2021. "Overdose Deaths Reached Record High as the Pandemic Spread." *The New York Times*, November 17.

Renn, Ortwin and Klinke, Andreas. 2004. "Systematic risk: a new challenge for risk management." *Science & Society*, 5(1): 41–46.

Robinson, Laura, Schulz, Jeremy, Ball, Christopher…Williams, Apryl. 2021. "Cascading crises: society in the age of Covid-19." *American Behavioral Scientist*, 65(12): 1608–1622.

Sanders, Shane. 2020. *The Economic Reason: A Piecemeal Guide to Your Inner Homo Economicus*. New York: Springer.

Shin, Jaeweon, Price, Michael, Wolpert, David, Shimao, Hajime, Tracey, Brendan, and Kohler, Timothy. 2020. "Scale and information-processing threshold in Holocene social evolution." *Nature Communications*, 11: 2394.

Sovacool, Benjamin, Del Rio, Dylan, and Griffiths, Steve. 2020. "Contextualizing the Covid-19 pandemic for a carbon-constrained world: insights for sustainability transitions, energy justice, and research methodology." *Energy Research & Social Science*, 68: 101701.

Taylor, Derrick. 2021. "A Timeline of the Coronavirus Pandemic." *The New York Times*, March 17.

Turchin, Peter. 2016. *Ages of Discord*. Chaplin, CT: Beresta Books.

Turchin, Peter. 2003. *Historical Dynamics*. Princeton, NJ: Princeton University Press.

Vasilyeva, Nadya and Ayala-Lopez, Saray. 2019. "Structural thinking and epistemic injustice," in Sherman and Goguen (Eds.), *Overcoming Epistemic Injustice: Social and Psychological Perspectives*. New York: Roman & Littlefield.

Weissboard, Richard, Batanova, Milena, Lovison, Virginia, and Torres, Eric. 2021. *Loneliness in America: How the Pandemic Has Deepened an Epidemic of Loneliness and What We Can Do About It*. Cambridge, MA: Harvard University, Graduate School of Education.

Worobey, Michael. 2021. "Dissecting the early Covid-19 cases in Wuhan." *Science*, 374(6572): 1202–1204.

Yang, Longqi, Holtz, David, Jaffe, Sonia…Teevan, Jaime. 2021. "The effects of remote work on collaboration among information workers." *Nature Human Behavior*, 6: 43–54.

Yin, Ling, Zhang, Hao, Li, Yuan...Mei, Shujiang. 2021. "A data driven agent-based model that recommends non-pharmaceutical interventions to suppress Coronavirus disease 2019 resurgence in megacities." *Journal of the Royal Society Interface*, 18, 2021.0112.

Yong, Ed. 2020. "How the pandemic defeated America." *The Atlantic*, September, 41–47.

Zamudio, Maria. 2020. "Covid-19 Deaths Are Rising in Chicago and Black Residents Remain the Most Likely to Die." *WBEZ Chicago*, November 20.

PART I

Health, economic, and environmental crises

2

CORONAVIRUS PANDEMIC

Chapter learning objectives

After reading this chapter, you will be able to:

LO1   Describe recurring losses during the coronavirus pandemic.
LO2   Discuss the birth of the coronavirus pandemic.
LO3   Explain characteristics of the coronavirus pandemic.
LO4   Consider the importance of global networks.
LO5   Analyze country outcomes.
LO6   Identify inequities.

Chapter outline

Recurring loss
Birth of a global crisis
Characteristics of the coronavirus pandemic
Network effects
Country outcomes
Inequities
Summary

Recurring loss

The contrast could not be sharper. More than a year after the coronavirus pandemic hit South America, death rates rose to levels that were among the highest in the world. In contrast, in the world's wealthiest countries, vaccinations were spreading, Covid-19 cases were easing, economies were recovering, and guarded

DOI: 10.4324/9781003310075-3

optimism was expanding. In South America, the spread of a new variant, Gamma, threatened to reverse gains in public health and economic systems. As the death toll rose, many communities in Brazil had to cut new graves in forests or abandon bodies on sidewalks. During this time, South America, with 8 percent of the world's population, accounted for 35 percent of the coronavirus deaths (Turkewitz and Taj, 2021). After a year of recurring loss, the surge in infections proved to be deadly.

In South America, why were the outcomes severe? First, ineffective healthcare systems, limited supplies of vaccines, and fragile economies complicated government responses. Second, because of large informal sectors and the inability to provide public assistance, it was difficult to maintain stay-at-home measures. But the third problem, longstanding inequality, established ongoing challenges. During the pandemic, millions of people lost income. But the working poor had to maintain employment however they could, even in the presence of the virus. During this time, protestors vented their anger about shutdown interventions, defied stay-at-home orders, and spread the virus. As a result, South America became "one of the globe's longest-haul Covid patients," creating instability with respect to both health and economic outcomes (Turkewitz and Taj, 2021).

On the other side of the world, India, a country with 1.4 billion people, experienced a devastating infection wave. During May 2021, more than 400,000 new daily cases and 4,500 deaths occurred, figures that most likely undercounted the actual totals. Hospitals were full. Before they had a chance to see doctors, many patients were dying in hallways. India was setting new morbidity and mortality records, putting the country in emergency mode. By the end of 2020, household lockdowns were successful in slowing the spread of the virus. Death counts were falling. However, because of pandemic fatigue and a lack of effective leadership, millions of people stopped taking precautions. In addition, an insidious new variant, Delta, different than the variant in South America, pushed India into unchartered territory. A record increase in daily cases, an inadequate healthcare response, and a shortage of vaccinations, despite the country's position as a leader in vaccination production, prolonged the crisis:

> India's dire needs (were) already having ripple effects across the world, especially for poorer countries. It had planned to ship out millions of doses; now, given the country's stark vaccination shortfall, exports have essentially been shut down, leaving other nations with far fewer doses than they had expected.
>
> *(Gettleman et al., 2021)*

In the presence of excess demand for vaccinations and an overwhelmed health care system, Delta infections accelerated.

Divergence

The contrast—infections and deaths falling in high-income countries in Europe and North America, plus Australia, New Zealand, and parts of Asia, but infections and deaths rising in South America, India, and other poorer regions—was not supposed to happen. A year into the pandemic, a global initiative, Covid-19 Vaccines Global Access (Covax), convinced 192 countries to promote access to vaccines, establish contracts with manufacturers for 2 billion doses, and coordinate a global distribution network. The program leveraged the resources of developed countries, establishing a framework to pay upfront costs. Then, Covax would help developing countries vaccinate their populations (Interlandi, 2021).

However, the inability of developed countries to complete the plan led to a surge in cases in South America, India, and elsewhere. Developed countries bypassed Covax, established contracts with vaccine manufacturers, and purchased the shots in the market. As a result, Covax lacked purchasing power; it could not compete with rich countries in securing contracts for vaccines. By the time Covax proved its viability, it was at the end of a line of countries ordering vaccines. In the developed world, vaccinations aided the process of recovery, but many low-income countries were not able to administer a single dose (Hall et al., 2021; Interlandi, 2021).

Ongoing costs

For all the benefits of vaccinations, ongoing costs existed. Manufacturing sites were not distributed throughout the world. Resource inputs and equipment were inefficient to meet production requirements. As the virus mutated, the world needed to ramp up manufacturing and deploy vaccinations, but global inequities established obstacles. By the time the world distributed 4 billion vaccine doses, 80 percent went to high- and upper-middle-income countries (Interlandi, 2021). By the time the world reached 12 billion doses, enough to provide two shots to more than 70 percent of the global population, very few doses reached the poorest countries (Interlandi, 2021). As the United States, the United Kingdom, and other developed countries attempted to vaccinate their entire populations, many low-income countries could not give initial doses. But developing countries that secured vaccinations experienced inefficient distribution networks. Even when they were successful in procuring supplies, operational failures often left stockpiles of doses moving toward expiration. In South America and India, weak healthcare systems, inadequate supplies of vaccinations, and a lack of political will led to new infection waves. The "campaign to vaccinate the world (was) floundering" (Mueller, 2021).

In this context, McKinsey & Company identified five factors that determined the success of vaccination campaigns: robust and efficient nerve centers that drove specific goals; efficient processes of delivery that determined vaccination rollouts; agile strategies that adapted to dynamic market conditions; past experiences that leveraged previous immunization campaigns; and programs that built hospital capacity. Together, these factors helped countries disseminate vaccinations; however, the factors were lacking in low-income countries (Hall et al., 2021).

Balance

Scientists identify a precarious balance between public health, technology, and economic development. Even though these forces create methods to fight a pandemic, they also contribute to disease outbreaks. In effect, disease outbreaks are similar to climate change in that they present a critique of the modern way of life, global economy, and methods of human organization. According to Adam Tooze (2021), who establishes a context for the coronavirus pandemic:

> Our use of resources across the globe, relentless incursions into the remaining wilderness, the industrial farming of pigs and chickens, our giant conurbations, the extraordinary global mobility of the jet age, the profligate, commercially motivated use of antibiotics, the irresponsible circulation of fake news about vaccines—all these forces combined to create a disease environment that was not safer, but increasingly dangerous.

While the elements of the modern world have been present for generations, the modern era is experiencing an increase in the threat of global disease outbreaks.

Chapter thesis and organization

As the novel coronavirus spread, altering human behavior, public health, and economic activity, it disproportionately impacted the most vulnerable members of society. Because race, ethnicity, and socioeconomic status affected health outcomes, discrimination eroded the health and well-being of poor communities. That is, health status correlated with community resources. But uneven outcomes occurred in all countries, including the United Kingdom and the United States. To address these concepts, the chapter discusses the birth of a global crisis, characteristics of the coronavirus pandemic, network effects, country outcomes, and inequities.

Birth of a global crisis

On December 29, 2019, samples were taken from a patient in Wuhan, China with a mysterious pneumonia, soon identified as a novel coronavirus. The United States and the United Kingdom reported their first cases on January 20 and January 29, respectively. By January 30, 9,976 cases were reported in 21 countries (Holshue et al., 2020). Local shutdown decisions accumulated into government-mandated lockdowns. However, even when

> governments did take the initiative, the efficacy of the measures put in place depended in large part on the active compliance of citizens, businesses, and organization, for whom government instructions served as a means to coordinate and rationalize their own responses.
>
> *(Tooze, 2021)*

Ed Yong (2020), writing in *The Atlantic*, questions how a virus a thousand times smaller than a dust mote "humbled and humiliated" the United States. The same question may apply to several countries that failed to control the virus. With the benefit of hindsight, it serves as an important but nuanced issue. Many factors determined the trajectory of the crisis, including the infectiousness of the virus, leadership, and collective action. For perspective, it is important to analyze the global epidemic curve, which demonstrates the shape of new daily cases.

Global epidemic curve

Between 2020 and 2022, the global epidemic curve of new daily infections demonstrates three infection waves followed by a much larger infection wave (Figure 2.1).

During the first year of the pandemic, new cases in North America and Europe fueled the initial infection wave. Because they did not experience the same level of connection, countries in South America and Africa initially experienced lower infection rates. When the virus spread, however, healthcare systems around the world struggled to maintain capacity. Governments experienced pressure to help businesses and households, while establishing safety guidelines. Economies contracted.

During the second year, a stuttered rollout of the global vaccination campaign disrupted the process of recovery. Three problems persisted: the emergence of new variants, reservoirs of untamed infections, and pandemic fatigue. In South

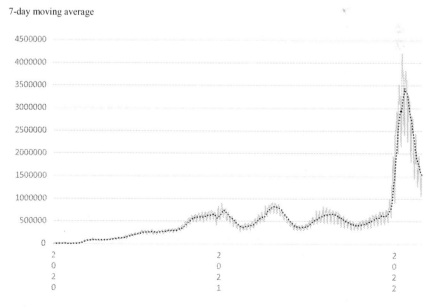

FIGURE 2.1 Global epidemic curve of new daily infections, 2020–2022.
Source: Our World in Data, https://ourworldindata.org/coronavirus

America and India, coronavirus transmission fueled the second and third infection waves. At the time, the world was creating hundreds of millions of vaccinations, but they were not widely distributed outside of the developed world. Spared the worst of the pandemic's effects during the first infection wave, many developing countries during the second and third waves could not implement or enforce sheltering-in-place, social distancing, or masking policies.

Mortality statistics illustrate the catastrophe. In May 2021, more than one year into the pandemic, 154 million global infections led to more than 3 million deaths. In March 2022, more than 2 years into the pandemic, 435 million global infections led to more than 6 million deaths. During the word's fourth infection wave, new daily infections peaked at 4,205,454 on January 19, 2022. According to the World Health Organization (WHO), deaths from Covid-19 were likely two to three times larger than official statistics. The discrepancy between the estimates and official data

> underscores the limited capacity of many countries to test their populations for the coronavirus and other weaknesses in official health data. For example, some Covid victims had died before being tested and their deaths did not appear in official reporting.
>
> *(Cumming-Bruce, 2021)*

By March 2022, 12–18 million deaths from Covid-19 likely occurred.

SEIR model

Epidemiologists, the researchers who study the incidence, distribution, and control of diseases, use the susceptible, exposed, infected, and recovered (SEIR) model. In a compartmental framework, the model establishes four categories for individuals (Figure 2.2).

Susceptibility and exposure

With a novel coronavirus, everyone is potentially susceptible. Early in a disease outbreak, with little public knowledge, a small number of infected people expose many others. During the coronavirus pandemic, Lawrence Wright (2021) argues

FIGURE 2.2 SEIR model.
Source: Author.

that three moments existed when events could have changed the trajectory of the crisis. He focuses on the United States, but his argument is relevant for other countries that suffered during the first year of the pandemic.

The first moment

The first moment that could have changed the trajectory of the pandemic, according to Wright (2021), occurred on January 3, 2020, when the director of the U.S. Centers for Disease Control and Prevention (CDC), Robert Redfield, in a conversation with his Chinese counterpart, George Fu Gao, the head of the Chinese Center for Disease Control and Prevention, discussed a report that an unexplained respiratory virus was spreading in Wuhan, China.

While both directors fretted the potential risk of a respiratory illness like the influenza pandemic of 1918 that killed more than 50 million people with more than 500 million infections, Gao initially argued that a lack of evidence existed for human-to-human transmission. While Redfield offered to send a medical team to analyze the outbreak, Gao did not have authorization from the Chinese government for this level of intervention. Without definitive intelligence, the U.S. public health contingent was not ready to recommend lockdowns, travel bans, or economic shutdowns. The problem was that, at the time, the virus was already spreading in human transmission networks in Wuhan and the Hubei Province, unknown to healthcare officials.

Realization

Addressing this situation, Redfield and Anthony Fauci, director of the National Institute of Allergy and Infectious Diseases, initially discounted the possibility that the virus could transmit asymptomatically in the absence of symptoms. They soon changed their minds. But, at the time, individuals infected with the virus were already exposing thousands of others throughout China. Even more, the virus spread through airline travel to Italy, Japan, South Korea, the United States, and elsewhere.

In early January 2020, if health officials in China and the United States had warned the world about the severity of the problem, they could have sounded an alarm, pushed for widespread travel bans, and implemented strong testing programs. But, with rising infections, it was soon clear that the outbreak was more like the 1918 influenza pandemic—which spread around the world—than the 2003 Severe Acute Respiratory Syndrome (SARS-Cov-1) outbreak—which also originated in China, and spread to several countries, but was largely contained.

Infections

After China's initial cases, infections varied. While it took 30 days for Spain to reach 100 confirmed cases, the United Kingdom took 34 days, and the United

States took 42 days (Brahma et al., 2020). The speed with which a virus spreads depends on the **basic reproduction number**, R_0 (R-naught), the average number of people an infected individual will infect. "The value of R_0 . . . is an important parameter that influences how quickly infections spread, and thus how quickly the number of hospitalized patients and fatalities will grow" (Chakraborty and Shaw, 2020). If $R_0 < 1$, an outbreak will cease. If $R_0 = 1$, the outbreak is **endemic**, maintained at a baseline level. When $R_0 > 1$, the outbreak may become a national epidemic or global pandemic. Suppose $R_0 = 2$. In this scenario, 1,000 infected people infect 2,000 others, who infect 4,000, and then 8,000, and 16,000, and so forth, leading to exponential growth. During the first year of the pandemic, R_0 was estimated to be between 2 and 2.5, depending on the region. The key in identifying infections is the implementation of testing systems. Because of the nature of SARS-Cov-2, testing served as a crucial element in identifying and stopping the spread of the virus. Without an effective testing mechanism, it is impossible to determine R_0 and the virus' level of infectiousness.

The second moment

The second moment that could have changed the trajectory of the pandemic was the "testing fiasco" that failed to identify early infections (Wright, 2021). Some countries implemented effective testing systems (Broom, 2020), but those that did not suffered during the initial infection wave. To complicate the matter, asymptomatic transmissions were rising. Many individuals were not aware of their infection status. Early in the pandemic, medical professionals often advised testing 5 to 7 days after exposure. Later, medical professionals learned to recommend testing 2 to 4 days after exposure (Anthes and Corum, 2022).

Asymptomatic cases

The reason asymptomatic individuals could be infectious involves the difference between the **incubation period** and **latency period** (Figure 2.3). The incubation period is the time between an individual becoming infected and showing

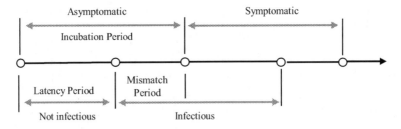

FIGURE 2.3 Periods in the course of an infection.
Source: Adapted from Christakis (2020), Figure 5, p. 49.

symptoms, averaging 5 days during the coronavirus pandemic. The latency period is the time between an individual becoming infected and being able to
spread the virus. During the latency period, newly infected individuals cannot
infect others. During the coronavirus pandemic, the latency period averaged 2
days.

The fact that the average latency period was shorter than the average incubation period served as the reason why the novel coronavirus spread through
the population. The difference, the **mismatch period**, occurred when infected individuals were asymptomatic (and did not know they were exposed)
but could infect others. As a result, a meaningful percentage of infected
individuals could spread the virus for 3 days on average before they were
symptomatic and showed signs of sickness (Du et al., 2020). (As a fascinating
comparison, for SARS-Cov-1, in 2003, the opposite held true: the average
latency period was longer than the average incubation period. This was one
of the reasons that SARS-Cov-1 did not spread through human transmission
networks as rapidly as SARS-Cov-2.)

Omicron

Nearly 2 years into the crisis, another variant, Omicron, spread through the
population. First appearing in South Africa, scientists could not initially forecast
how it would behave. The problem was that Omicron had a distinctive combination of 50 genetic mutations. The variant drove new infections to record levels,
disrupted economic activity, and accounted for almost 100 percent of new cases.
The fourth global infection wave, during the winter of 2021–2022, resulted from
the Omicron variant. For perspective, the Delta variant had an incubation period that averaged 4 days. With Omicron, the incubation period averaged 3 days
(Figure 2.4).

However, compared with Delta, the Omicron variant caused milder disease
effects. It inflicted the upper airway, including the nose and the throat, as opposed to the lungs, which occurred with Delta. Over time, Omicron infected
people who were vaccinated or recovered from infections, but hospitalizations
did not increase proportionately. Compared with other variants, Omicron was
less likely to cause severe illness (Zimmer and Ghorayshi, 2021).

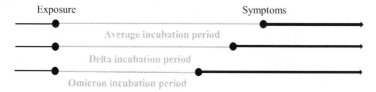

FIGURE 2.4   Incubation periods.
Source: Adapted from Anthes and Corum (2022).

However, healthcare systems struggled with the consequences of Omicron, as the elderly battled new infections. Individuals without vaccinations or boosters and no immunity to the virus suffered more from the exposure to Omicron. Two years into the pandemic, the United Kingdom, with ten times the population of the state of Massachusetts, experienced deaths from Covid-19 at half the rate of Massachusetts. The reason: individuals in Massachusetts were less likely to have vaccinations and booster shots (Hanage, 2022).

Viral load

Viral load exists as a measure of the amount of a virus that accumulates in the body. When the viral load is high, the degree of infectiousness rises. During the pandemic, individuals reached their peak viral loads 3 days after infection and were clear 6 days after that. While Omicron and Delta led to similar results with respect to viral load, Omicron caused less severe cases of disease. In general, as disease outbreaks progress, this result is common. Compared with Delta, individuals with Omicron were less likely to go to hospitals, require intensive care, or need mechanical ventilation. But the relative mildness of Omicron stemmed from the reality that it infected more individuals. Omicron often evaded antibodies produced in the body after vaccination, creating breakthrough cases. After the emergence of symptoms, victims of the Omicron variant recovered, required hospitalization, or experienced **long-haul Covid**, but they were recommended isolation (Figure 2.5).

Long-haul Covid

The problem of long-haul Covid, the reality that some victims of the disease for long periods of time experienced stubborn and persistent fatigue, shortness of breath, insomnia, and loss of taste or memory, revealed that the illness would not recede (George, 2021). Long-haul symptoms ranged from weeks to months. An article in *The New England Journal of Medicine* argued that the symptoms affected "organ systems, occur(ed) in diverse patterns, and frequently (got) worse after physical or mental activity" (Phillips and Williams, 2021). But the article explained that, given the history of post-infection syndromes, long-haul Covid

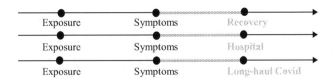

FIGURE 2.5 Outcomes of Omicron.
Source: Adapted from Anthes and Corum (2022).

was not surprising. Because long-haul symptoms occurred in many of those who were infected, however, it existed as a stubborn and persistent characteristic of the coronavirus crisis.

Isolation

The CDC originally recommended that individuals who tested positive for Covid-19 should isolate for 10 days. But observations led to new guidelines: individuals, if they were asymptomatic or experienced a few symptoms, should isolate for 5 days (Figure 2.6). In the latter case, individuals were supposed to wear masks for an additional 5 days when they were in public, a recommendation that many did not follow (Anthes and Corum, 2022).

Recovery

At the end of the fourth global infection wave, cases decreased while immunity against Covid-19 increased. William Hanage (2022) of the Harvard T.H. Chan School of Public Health explained:

> Every exposure, whether to the virus or vaccine, reduces the likeli-hood of severe illness or subsequent ones. That's because each time our immune systems "see" the spike proteins on the outside of the coro-navirus, which is the target for all the vaccines in use, they get better at responding to them. Infections get less severe, on average, over time not just because the virus is changing but also because our bodies are getting better at handling it.

But this pattern did not guarantee repetition. The immunity from Omicron did not end the pandemic. "Instead of the virus going away, the nature of the disease it causes changes to a point that people consider it a tolerable risk" (Hanage, 2022).

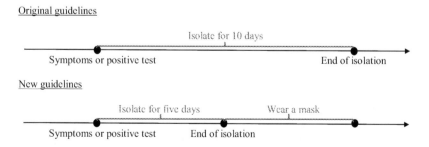

FIGURE 2.6 Guidelines for isolation.
Source: Adapted from Anthes and Corum (2022).

Characteristics of the coronavirus pandemic

Why did the coronavirus pandemic create so much damage? A number of reasons existed, including aerosols, antipathy to globalism, time necessary to reach **herd immunity** (the protection that occurs when a sufficient percentage of the population becomes immune), uneven leadership, roadblocks, and problems with methods of intervention.

Aerosols

Throughout 2020, the WHO argued that the main way the virus spread was through respiratory droplets, expelled to surfaces. But the revised reason, announced in December 2020, was that coronavirus transmission occurred with aerosols that remained suspended in the air. As a result, the U.S. CDC updated its guidelines. But this shift challenged "key infection control assumptions . . . putting a lot of what went wrong (during the pandemic) in perspective" (Tufekci, 2021). If the world had known that aerosols served as the threat to public health, leaders could have emphasized the importance of outdoor rather than indoor gatherings, masks, ventilation and air filters, and prevention of super-spreading events in airports, nursing homes, and meatpacking plants.

Antipathy to globalism

Less investment in national and local health security, according to Richard Horton (2020), editor of the medical journal *The Lancet*, in his book on *The Covid-19 Catastrophe*, also played a role, reflecting a general "antipathy to globalism." Globalism, an appreciation of the cooperation, interdependence, and solidarity between people and nations, was lacking among world leaders such as Donald Trump in the United States, Narendra Modi in India, and Jair Bolsonaro in Brazil. Brexit, the withdrawal of the United Kingdom from the European Union, which became official at the beginning of 2021, 1 year into the pandemic, existed as a nationalist policy. But the problem with anti-globalism is that, because of global networks, the SARS-Cov-2 outbreak quickly became a pandemic, and thus required cooperation between countries, not nationalist responses.

Herd immunity

The most effective way to end a pandemic is to have an antiviral drug or vaccine that inhibits the virus, leading to herd immunity. When a population reaches herd immunity, the pandemic ends: the basic reproduction number falls below one. However, if a country requires a higher percentage of the population to be in the recovered compartment for herd immunity, it takes longer for the pandemic to end. Mathematically, when the proportion of the population in the recovered compartment in the SEIR model exceeds $1 - 1/R_o$, the population achieves herd immunity. For example, if the basic reproduction number R_0 = 2, herd immunity is reached when more than 50 percent of the population recovers through vaccination, antiviral drugs, or surviving infection. If R_0 = 3, a population reaches herd immunity when more than 67 percent of the population has recovered. As a result, maintaining a low basic reproduction number and implementing an effective and equitable vaccination program constitute important policy goals.

Antiviral drugs

Antiviral drugs require virus identification in order to interfere with the ability of the virus to replicate in human cells. Modern technologies revolutionized the way in which scientists identify viruses. In 2003, with SARS-Cov-1, scientists identified the virus 6 months after initial reports of disease. In 2020, with SARS-Cov-2, scientists identified the virus one month after initial reports of disease. Because of the development of new methods to isolate viruses and sequence their genomes, rapid virus identification occurs. Antiviral drugs work by blocking one or more steps in the viral lifecycle: viral entry into human cells, replication, assembly, and release into the bloodstream (Chakraborty and Shaw, 2020). In October 2020, less than 1 year into the pandemic, the first antiviral drug was available to treat patients with Covid-19.

Vaccinations

Vaccine development began, in January 2020, soon after the identification of the virus. Vaccine dissemination began in December 2020, a fast timeline in historical perspective. In the context of the SEIR framework, the vaccines created a pathway from susceptible to recovered individuals. But, during the coronavirus pandemic, problems with vaccinations persisted. An ideal model of the process, distributed manufacturing, decouples vaccine manufacturing and allocation. That way, countries with comparative advantage in manufacturing would produce the shots. The system would then deploy the shots to the areas where they are needed. The advantages of distributed manufacturing, when regions around the world have modern vaccination hubs, include the ability to respond to local needs, preparedness, and stopping outbreaks before they escalate. The reason this ideal model is important is that, in the absence of an international agreement, standard, or regulating agency, countries that export doses have fewer doses available for domestic consumption. For example, during the coronavirus crisis, the European Union, the United States, and India established export controls on vaccinations that enabled them to prioritize domestic consumption (Interlandi, 2021). In addition, barriers to manufacturing complicated the process. Companies had to build or upgrade manufacturing plants. Even more, for the creation of vaccines, knowledge of the process was not evenly distributed. Vaccine makers could refuse to share their technology, even when they received public subsidies. The mRNA shots distributed during the coronavirus pandemic required hundreds of ingredients and specialized equipment, which were not available in under-sourced environments. The development of vaccines required multiple countries, companies, and global supply chains. For an efficient system, a coordinator was needed to establish production goals, resource allocation, and future planning.

> If countries commit to a global vision for vaccination and if they work together toward its realization, it's possible the vast inequities . . . will be avoided. . . . If individualism is allowed to prevail instead, the world's resources will only grow more concentrated, and the world's poorest nations will continue to be left out.
>
> *(Interlandi, 2021)*

Leadership

During the initial months of the coronavirus pandemic, leaders in the United States and the United Kingdom were complacent. They underestimated the problem: the risk that a new pathogen could ravage healthcare networks and economies was not initially in the realm of possibilities. But, during a global pandemic, governments exist as the first line of defense. Leaders must assess risk, assume responsibility, mobilize resources, adjust to changing circumstances, and determine future pathways. In a global perspective, during the first year of the pandemic, some national leaders implemented interventions that led to the eradication of disease outbreaks, other leaders contained them, and a third category failed to achieve either outcome.

One area of research addressed why countries led by women seemed to be more successful in fighting the coronavirus (Aldrich and Lotito, 2020). During the first year of the pandemic, Jacinda Ardern of New Zealand, Angela Merkel of Germany, Sanna Marin of Finland, and Tsai Ing-wen of Taiwan presided over efforts to contain the virus and implement systems of testing and contact tracing. The idea was that the success of these leaders was due to their effective leadership skills. In addition, countries with women leaders had inclusive values, the ability to consider different perspectives, and long-term perspectives (Taub, 2020). Another area of research addressed the reality that the provision of effective healthcare responses involved humanitarian operations and the consideration of vulnerable populations (Sokat and Altay, 2021). Thus, leaders assumed an important role in motivating the public to take collective action, cooperate with public health measures, and help marginalized members of society.

Roadblocks

During the coronavirus pandemic, countries experienced economic, political, and social roadblocks to recovery, including the prioritization of economic activity over public health, pandemic fatigue, uncoordinated policy interventions, and the anti-vaccination status of some members of the population. In Brazil, India, and the United States, these factors prolonged the pandemic; however, the reluctance of some individuals to seek vaccinations served as an important roadblock. According to Dr. Sema Sgaier (2021) of the Harvard T.H. Chan School of Public Health, for four reasons, some individuals refused vaccinations. They were watchful—waiting to see if vaccinations would be effective—concerned about the time or cost in getting vaccinated, distrustful of healthcare systems, or skeptical about the severity of the crisis.

Methods of intervention

Attempting to slow the spread of disease, two methods of intervention exist: pharmaceutical and nonpharmaceutical. **Pharmaceutical interventions**, such as vaccinations and antiviral drugs, employ the medical establishment. While vaccine development during the coronavirus pandemic occurred in record time, the process did not make vaccines available until December 2020, almost a year after the start of the pandemic. During early infection waves, the application of medications was hampered by the number of sick patients. As a result, **nonpharmaceutical interventions**, the health interventions that are not based on medication, played an important role. Before the dissemination of vaccinations, nonpharmaceutical interventions flatten the epidemic curve, which means delaying the peak number of infections and providing the healthcare system with the time necessary to prepare for the onslaught of infections. But problems exist. First, when hospitals are overwhelmed with new cases, nonpharmaceutical interventions are not successful in flattening the epidemic curve. Writing in *Scientific American*, Katherine Courage (2021) documents that, when intensive care units experience an overwhelming case load and an insufficient number of doctors and nurses to handle the surge, resources become scarce, patients do not receive sufficient care, and stress and anxiety rise among patients and staff. Second, rising costs accompany nonpharmaceutical interventions. During lockdown, when individuals cannot leave their homes, tradeoffs occur, including a rise in anxiety, domestic violence, isolation, mental health problems, and quarantine fatigue. Nonpharmaceutical interventions exist in two categories: individual and collective. Not mutually exclusive, these methods work in a complementary framework.

Individual and collective measures

Individual measures include mask-wearing, social distancing, and staying home when sick. Because these actions involve personal choice, individuals have flexibility. Collective measures modify human interaction and are implemented and managed by government, including business shutdowns, border closings, contact tracing, quarantines, school closings, stay-at-home orders, and testing. Because collective interventions impose restrictions on both infected and non-infected individuals, they are subject to pushback. As a result, they may lead to resistance and resentment, depending on the economic, political, and social contexts. In the United States and Brazil, several reasons for pushback occurred, including political division, misinformation, and personal liberty.

The third moment

After the first moment of not sufficiently warning the public about the threat of the virus and the second moment of the inability to implement effective testing mechanisms, the third moment that could have changed the trajectory of the pandemic in the United States, according to Wright (2021), related to masks. During the early months of the coronavirus pandemic, the failure was the choice at the highest level of government not to establish a practical, efficient, and nation-wide mandate for masks, even after it became clear that masks were effective in helping to slow the spread of the coronavirus. While East Asian countries have a culture of mask-wearing, the United States does not. For context, 6 months into the pandemic, more than 100 countries had issued nation-wide mask mandates, but not the United States. In a study that addressed variation in mortality for Covid-19 across 196 countries, Leffler et al. (2020) found that colder average country temperatures, duration of the outbreak, proportion of the population over 60 years of age, prevalence of smoking, and urbanization were correlated positively with higher mortality rates, but mask-wearing was correlated negatively with mortality rates.

Network effects

In *The Coming Plague*, Laurie Garrett (1994) argues that

> humanity will have to change its perspective on its place in Earth's ecology if the species hopes to stave off or survive the next plague. Rapid globalization of human niches requires that human beings everywhere on the planet go beyond viewing their neighborhoods, provinces, countries, or hemispheres as the sum total of their personal ecospheres.

In a world of global networks, systems of interconnection guide communication and exchange. But global interconnection increases our level of vulnerability to infectious diseases.

Globalization

Globalization, the interconnection of the world's people through all forms of exchange, facilitates economic activity. Urbanization, industrialization, and modernization lead to global networks of exchange, including trade, technology, finance, information, culture, and tourism. Globalization increases mobility. But trade and travel spread infections. Urban growth, accelerated by rural-to-urban migration, drives infectious diseases through commuting, population density, and human transmission. Countries with increasing human interaction (domestic plus foreign) establish the conditions for pandemics (Antrás et al., 2020). Throughout history, pandemics were closely intertwined with globalization. In the fourteenth century, trade routes spread the bubonic plague from China to Europe. During World War I, the influenza pandemic, following armies, led to millions of deaths. The Asian flu of 1957, reported in 20 countries, spread by both land and sea. In January of 2020, an individual from Wuhan who traveled to Germany initiated an early person-to-person spread of the novel coronavirus. But these examples are not unique. Countless cases link travel with contagion. Integration creates material gains and a network for disease transmission.

Negative globalization

The impact of globalization depends on the context. When a company sends jobs overseas, gains in employment in one area are offset by job losses in another. Globalization normally leads to freer trade arrangements, but barriers remain. To avoid paying higher tax rates, multinational corporations exploit tax havens. Unfair working conditions, pollution flows, and ecological damage characterize global economic arrangements. **Negative globalization** includes the networks of transmission and structures that amplify these problems. Borderless diseases, such as malaria, plague, and tuberculosis, have global reach. In the current century, SARS-Cov-1 in 2003, Middle East Respiratory Syndrome in 2012, and SARS-Cov-2 in 2020 serve as examples. These pathogens, which cause deadly diseases, evade border controls and travel restrictions.

Impact of the pandemic on globalization

The coronavirus pandemic placed an unprecedented burden on the global economy. During 2020, the global economy contracted by more than 4 percent. In national economies, consumers spent less on goods, decreasing aggregate demand. Government interventions reduced the production of output, decreasing aggregate supply. The economic fallout included lower trade volumes, event cancelations, and decreased workforce participation. In disease hotspots, healthcare systems experienced equipment and staff shortages. Because of severed transportation networks, cargo ships were refused entry into port, manufacturing plants were shuttered, and global supply chains were fractured.

Country outcomes

During the coronavirus pandemic, countries experienced different infection waves. Several factors influenced these patterns, including leadership, roadblocks, and the distribution of vaccinations. Some countries quickly flattened the epidemic curve, creating an environment for recovery. Others did not. During the first year of the pandemic, the Boston Consulting Group categorized countries with respect to their actions to flatten the epidemic curve, including "crush and contain," "flatten and fight," and "sustain and support" (Gjaja et al., 2020).

Crush and contain

Countries that crush and contain, including China, New Zealand, and South Korea, implement rapid and stringent lockdowns, establish aggressive measures to flatten the epidemic curve, experience effective leadership, and do not suffer from roadblocks. With a population of 4.9 million, New Zealand's epidemic curve demonstrates an initial infection wave that reached peak daily cases of 89 on April 2, 2020 (Figure 2.7). By June 2020, New Zealand declared the pandemic over, reporting one of the lowest coronavirus-related mortality rates among the 37 Organization of Economic Cooperation and Development (OECD) countries. During the first 18 months of the pandemic, New Zealand experienced a small and manageable level of cases. But infection waves at the end of 2021 and at the beginning of 2022 pressured the country to implement new interventions to fight rising infections from the Omicron variant (McKenzie, 2022).

Flatten and fight

Countries that flatten and fight, the most common approach, including Australia, Germany, and the United Kingdom, implement nonpharmaceutical interventions until the arrival of a vaccine. These countries monitor infections

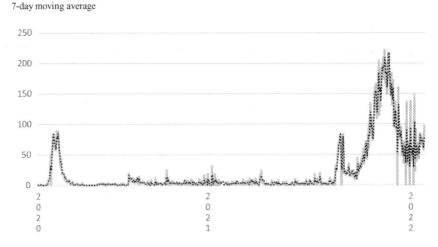

7-day moving average

FIGURE 2.7 New Zealand's epidemic curve of new daily infections, 2020–2022.
Source: Our World in Data, https://ourworldindata.org/coronavirus

7-day moving average

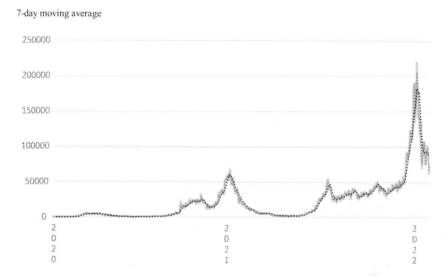

FIGURE 2.8   United Kingdom's epidemic curve of new daily infections, 2020–2022.
Source: Our World in Data, https://ourworldindata.org/coronavirus

with testing and contact tracing. During the spring of 2020, after the United
Kingdom's initial lockdown reduced social interaction, the government of Bo-
ris Johnson restored economic activity, but infections rose. When the country
exited its second lockdown, in December 2020, it eased restrictions by allowing
some economic activity and social interaction. During both periods, as hospitalizations, infections, and deaths increased, the government had to pause the
process of recovery (El-Erian, 2021). With a population of 66.65 million, the
United Kingdom's epidemic curve demonstrates a small infection wave, in April
2020, a second wave, in November 2020, a third wave, in January 2021, and then
several waves in 2021, culminating with a peak number of infections of 219,290
on January 4, 2022 (Figure 2.8).

Sustain and support

Sweden implemented an alternative plan: sustain and support. The idea was not
to flatten the epidemic curve but to implement voluntary and targeted restrictions for vulnerable members of the population, particularly the elderly, while
keeping most of the economy and society open. Interventions received high
levels of compliance, including the closure of schools and universities. But the
Swedish response existed as an outlier, associated with a greater loss of economic
activity compared with similar countries. Sweden resisted lockdowns for much
of the first year, with no mandatory limits restricting crowds in commercial
districts, public transportation, or other gathering places. This herd immunity
approach, which resulted from Sweden's national pandemic strategy, led to a slow

7-day moving average

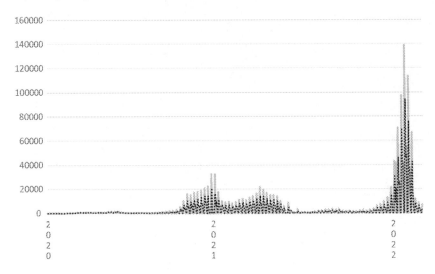

FIGURE 2.9 Sweden's epidemic curve of new daily infections, 2020–2022.
Source: Our World in Data, https://ourworldindata.org/coronavirus

but gradual increase in infections during the first 6 months of the pandemic. With a population of 10.23 million, Sweden's epidemic curve demonstrates an initial infection wave peaking in December 2020, a second wave peaking in April 2021, and the largest wave in January 2022 with a peak number of infections of 138,985 on January 25, 2022 (Figure 2.9).

Countries may struggle

The United States

In the United States, during the first 3 months of the pandemic, more people died of Covid-19 than were killed in action during a 10-year period of time (1965–1975) in the Vietnam War. By the end of 2021, in the United States, more people had died from Covid-19 than the number of U.S. soldiers killed in both world wars. With a population of 328 million, the country's epidemic curve demonstrates small infection waves in April and July 2020, a larger wave in December 2020, an additional wave in September 2021, and the largest infection wave in January 2022, with the peak number of new daily infections (1,268,167) on January 10 (Figure 2.10).

India

India, with a diverse population of 1.4 billion people—second only to China— and social and economic disparities, restricted travel during the pandemic and

7-day moving average

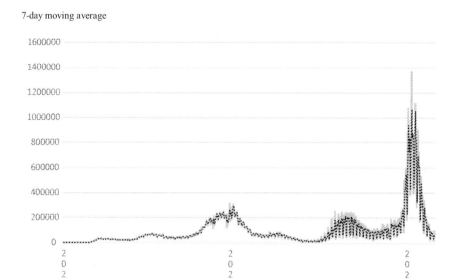

FIGURE 2.10 United States' epidemic curve of new daily infections, 2020–2022.
Source: Our World in Data, https://ourworldindata.org/coronavirus

7-day moving average

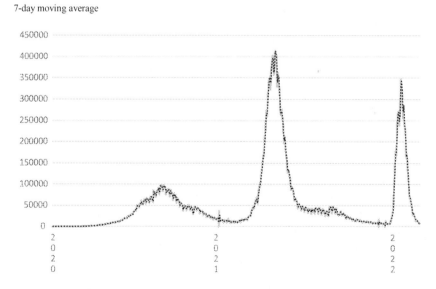

FIGURE 2.11 India's epidemic curve of new daily infections, 2020–2022.
Source: Our World in Data, https://ourworldindata.org/coronavirus

initially maintained a low mortality rate. However, problems emerged, including a lack of testing, shortages of healthcare workers, and infections spreading in slums in cities like Mumbai and Calcutta. The first infection wave occurred in September 2020 (Figure 2.11). In areas with overcrowded living conditions and inadequate sanitation, governing authorities struggled to maintain health standards and distribute vaccinations, despite the country's high level of vaccine production. The second wave in April and May of 2021 resulted in a peak number of infections of 414,188, on May 6. January 2022 brought a third infection wave.

Case study 2.1 The importance of trust

Even though the public sector implements rules and regulations, directives are not self-executing. Effective policy requires guidance and cooperation. Trust is essential for public health. Policymakers must establish constructive measures that balance costs and benefits, but social attitudes toward collective action help to determine outcomes. Society, for example, may value measures that benefit the common good, but not government control. This reality may lead to positive outcomes. In Demark, more than 2 years into the pandemic, "people (were) in favor of vaccines, with more than 81 percent of adults doubly vaccinated, but also very opposed to vaccine mandates" (Klein, 2022). Denmark minimized new infections but did not require a command-and-control approach from the government. A study published in *The Lancet* argues that, in the first 21 months of the coronavirus pandemic, the United States had 545 cases per 1,000 residents, the United Kingdom 374, South Korea 28, and Taiwan seven (Bollyky et al., 2022). With a large economy and advanced technology, why did the United States struggle? To answer the question, the study's researchers tested numerous factors for predictive power, including age, air pollution, cancer rates, exposure to previous coronaviruses, Gross Domestic Product (GDP), health insurance coverage, hospital beds per capita, pandemic preparedness, population density, and trust in fellow citizens and government. When they analyzed factors that predicted infections, many of the usual factors—exposure to previous coronaviruses, GDP, and population density—were not statistically significant. But trust in fellow citizens and the government was statistically significant. The study finds that "the level of trust is something that a government can prepare for and earn in a crisis, and our analysis suggests doing so may be crucial to mount a more effective response to future pandemic threats" (Bollyky et al., 2022). The conclusion is that, throughout the pandemic, higher levels of trust in fellow citizens and government led to a reduction in infections.

Inequities

An important consequence of the coronavirus pandemic was that it dispro-portionately impacted the most vulnerable members of society. The spread of Covid-19 overwhelmingly affected the poor and elderly. That is, as the disease progressed, a social gradient emerged. With the potential to infect all members of society, the novel coronavirus instead exploited and expanded every avenue of inequity. As a result, a salient feature of the coronavirus pandemic was the inequitable nature of human devastation.

Vulnerability

The theme that individuals experiencing unfavorable positions within the social order suffered the most during the coronavirus pandemic informs research on pandemic outcomes, especially health effects. As infections rise and economies contract, it is important to consider what institutions are failing and whom they are failing the most. The answer to the first question exists in the form of epi-demiological and economic data on rising infections and unemployment (Par-tington, 2021; Casselman, 2020). The answer to the second question relates to mobility: while wealthy households flee urban areas for country homes and much of the middle class experiences the benefit of remote work, policy interventions may fail to protect the working poor. The result: "pandemic precarity dispropor-tionately affects historically disadvantaged groups, widening inequality" (Perry et al., 2021).

Disparities

The coronavirus pandemic exposed disparities among ethnic minority groups and mothers. In these contexts, individual experiences did not depend on chance or fortune. Instead, they depended on race, class, gender, and socioeconomic status. Significant factors included access to healthcare resources, family respon-sibilities, inequalities, and pre-existing conditions such as diabetes and cardio-vascular disease (El-Khatib et al., 2020). During the pandemic, high-income households spent more time outdoors, avoided face-to-face interactions, and saved more of their disposable income. Before the pandemic ended, employ-ment among higher-paid professional workers returned to pre-pandemic levels, whereas many low-paid workers in the service sector suffered from higher levels of economic insecurity. As the beginning of the chapter explains, an inequitable global vaccine rollout in 2021 created similar problems, leaving lower-income households at a disadvantage. With respect to education, poorer households were less likely to provide adequate access to educational resources for their children, widening the learning gap. While life expectancy fell for everyone, it fell most for minority populations (Serkez, 2021). During the pandemic, rising morbidity, mortality, and unemployment were a function of access to healthcare services, age and coresidence, characteristics of the disease, economic disparities, welfare programs, gender roles, hygienic and sanitary conditions, income, minority status, patterns of migration and displacement, population density, poverty and discrimination, and systems of case notification. The following sections consider three of these factors: age and coresidence, gender roles, and minority status. The reader is encouraged to research the others.

Age and coresidence

Covid-19 impacted age groups in an uneven manner. In general, age is a marker of the gradual accumulation of experience over the course of a human lifetime. But age correlates with disabilities and chronic diseases. Although everyone was susceptible to Covid-19, it was primarily a disease of adulthood. The risk of mortality increased with age. As a result, care homes for the elderly faced difficult health environments with restricted visits and staff susceptible to infection. In addition to age, coresidence—multigenerational living arrangements—served as a risk factor. During the first year of the pandemic, the age structure of North American and Northern European countries increased the risk of Covid-related deaths. In African and Asian countries, higher levels of coresidence for elderly relatives increased the level of risk. But Southern European countries, such as Greece, Italy, and Spain, with older age structures and higher levels of coresidence, experienced the highest levels of risk. The implication was that school closings had a smaller impact than interventions aimed at older age groups (Esteve et al., 2020).

Gender

In many countries, women were disproportionately affected by the pandemic, especially during the initial infection wave when children experienced remote education. Working mothers had to juggle their professional responsibilities, household duties, and the supervision of their children's online education. In response, some mothers left the workforce. As a result, stress and anxiety rose (Grose, 2021). But public policy was slow to address the burden. Mothers were responsible for domestic work and childcare. Before the pandemic, in many countries, the labor force participation rate for women equaled or exceeded the rate for men. But the pandemic altered this trend. Short-term outcomes included an increase in depression, resignation, and exhaustion among mothers who had to coordinate childcare, education, and work. Long-term outcomes included lost employment or promotions, less retirement income, and emotional distress (Bennett, 2021).

Minority status

During the coronavirus pandemic, a pattern emerged: once sick, individuals in low-income quintiles, often members of minority populations, were more likely to die of Covid-19. The reasons for an inequality of outcomes included higher rates of chronic disease and less access to healthcare resources. By the end of April 2020, 2 months into the pandemic, Prince George County in Maryland, one of the wealthiest majority-black counties in the United States, reported some of the highest death tolls from Covid-19 and the most exposures in the Washington D.C. area.

These results corresponded to a growing trend that, compared with Black Americans, white Americans were less likely to die from Covid-19. Because many people in Prince George County were frontline workers who experienced daily exposures, the coronavirus ravaged the area. As a risk factor, members of the Black community experienced diabetes, hypertension, and obesity at rates higher than national averages. Because of fewer primary care doctors and hospital beds, residents of the county were less likely to get treated for the virus. Prince George's hospitals were inundated with Covid-19 patients, forcing the transfer of some to nearby facilities (Chason et al., 2020).

Another problem emerged. As people died from Covid-related illnesses, others suffered from unrelated afflictions. The reason was that, because hospital resources were dedicated to fighting Covid-19, many diseases were undiagnosed. Some individuals missed preventive care appointments, exacerbating problems from pre-existing conditions. In the United States, people of color were disproportionately affected. For white people, during 2020, excess deaths—the number of individuals who died from all causes, in excess of the expected number of deaths—increased by 11.9 percent. For African Americans, excess deaths increased by 32.9 percent. For the Latinx population, excess deaths increased by 53.6 percent (Rossen et al., 2020). Because of a lack of access to healthcare resources, this hidden crisis targeted minority populations. In this context, Dr. Wayne A. Frederick (2021), president of Howard University and professor of surgery at Howard University College of Medicine, argues:

> Too often, Black and brown patients are left behind. During the early stages of the pandemic, when we were reconfiguring our health care system to tackle Covid-19, we neglected to protect many of those working in our health care centers, grocery stores, meatpacking plants and public transit systems. We overlooked the custodial staffs who continued to ensure our businesses and medical offices were clean, sanitary and safe. When we canceled appointments and shut down screening centers, we forgot about those who faced other life-threatening conditions. By now returning our focus to the full spectrum of patients' needs, we can prevent more deaths and protect the communities of color that have endured an undue share of our national devastation.

Lessons

What lessons stemmed from the coronavirus pandemic? First, the early months of the crisis marked a significant change in global human behavior (Johnson, 2021). Vast segments of society shut down and then implemented measures to flatten the epidemic curve. At the time, it was not clear that such actions were possible. Second, the rapid identification of the SARS-Cov-2 virus about 20 days after the first reported outbreak and sequencing of the genome shortly thereafter represented scientific breakthroughs (Johnson, 2021). Third, the endemic pattern of a new disease is understood in a retrospective manner (Shaman, 2022). Outbreaks depended on population immunity and virus mutations. Fourth, while an optimistic forecast for SARS-Cov-2 was that it would settle into a less disruptive flulike pattern, the pessimistic scenario was that it would continue to generate infectious variants for years to come (Shaman, 2022). Fifth, the pandemic disproportionately impacted the most vulnerable members of society (Abedi et al., 2020). As a result, to reduce the risk of infection, resolve the problem of suboptimal healthcare, and improve the effectiveness of targeted interventions, it is important to evaluate the impact of the pandemic on the individuals who benefit the least from the existing order.

Summary

A global pandemic increases morbidity, mortality, and unemployment. The global epidemic curve demonstrates multiple infection waves. While vaccinations reduce both infections and deaths, the initial vaccination rollout benefits high-income countries at the expense of low-income countries. The SEIR model demonstrates that the rates of exposure, infection, and recovery determine the risk to society. During the development of vaccines, countries rely on non-pharmaceutical interventions to limit the spread of the virus. But a lack of public cooperation and roadblocks to recovery prolong the crisis. Overall, a global crisis disproportionately impacts the most vulnerable members of society, including the elderly, mothers, and members of minority populations.

Chapter takeaways

LO1   Inequality, ineffective healthcare systems, limited supplies of vaccinations, and fragile economies make it difficult for countries to recover from a global crisis.

LO2   Susceptible populations, persistent infections, and slow recoveries prolong a global pandemic.

LO3   The characteristics of the coronavirus pandemic include aerosols, antipathy to globalism, problems reaching herd immunity, uneven leadership, roadblocks, and interventions.

LO4   Global interconnection increases the risk of negative flows such as viruses.

LO5   Country outcomes relate to strategies of intervention: crush and contain, flatten and fight, and sustain and support.

LO6   A global pandemic disproportionately impacts the most vulnerable members of society, including the poor, elderly, mothers, and individuals with minority status.

Key terms

Basic reproduction number
Endemic
Herd immunity
Incubation period
Latency period
Long-haul Covid

Mismatch period
Negative globalization
Nonpharmaceutical interventions
Pharmaceutical interventions
Viral load

Questions

1   During the coronavirus pandemic, what were examples of recurring loss?
2   How does the SEIR model provide a framework to analyze the coronavirus pandemic?
3   Why is the difference between incubation and latency periods important?
4   What is the link between globalization and virus transmission?
5   Did the coronavirus pandemic have a lasting impact on globalization?
6   In controlling the spread of a virus, why are leaders important?
7   For an individual country, how would you characterize its infection waves?
8   During the coronavirus pandemic, what factors impacted morbidity and mortality?

References

Abedi, Vida, Olulana, Oluwaseyi, Avula, Venkatesh, Chaudbary, Durgesh, Khan, Ayesha, Shahjouei, Shima, Li, Jiang, and Zand, Ramin. 2020. "Racial, economic, and health inequality and Covid-19 infection in the United States." *Journal of Racial and Ethnic Health Disparities*, 8(3): 732–742.

Aldrich, Andrea and Lotito, Nicholas. 2020. "Pandemic performance: women leaders in the Covid-19 crisis." *Politics & Gender*, 16(4): 960–967.

Anthes, Emily and Corum, Jonathan. 2022. "Charting an Omicron Infection." *The New York Times*, January 22.

Antrás, Pol, Redding, Stephen, and Rossi-Hansberg, Esteban. 2020. "Globalization and Pandemics." NBER Working Paper 27840 (September): National Bureau of Economic Research.

Bennett, Jessica. 2021. "Three Mothers, on the Brink." *The New York Times*, February 7.

Bollyky, Thomas, Hulland, Erin, Barber, Ryan…Dielman, Joseph. 2022. "Pandemic preparedness and Covid-19: an exploratory analysis of infection and fatality rates, and contextual factors associated with preparedness in 177 countries, from Jan. 1, 2020, to Sept. 30, 2021." *The Lancet*, 399(10334): 1489–1512.

Brahma, Dweepobotee, Chakraborty, Sikim, and Menokee, Aradhika. 2020. "The Early Days of a Global Pandemic: A Timeline of Covid-19 Spread and Government Interventions." *Brookings*, April 2.

Broom, Douglas. 2020. "These Are the OECD Countries Testing Most for Covid-19." *World Economic Forum*, April 30.

Casselman, Ben. 2020. "U.S. Economy Stumbles as the Coronavirus Spreads Widely." *The New York Times*, November 25.

Chakraborty, Arup and Shaw, Andrey. 2020. *Viruses, Pandemics, and Immunity*. Cambridge, MA: The MIT Press.

Chason, Rachel, Wiggins, Ovetta, and Harden, John. 2020. "Covid-19 Is Ravaging One of the Country's Wealthiest Black Counties." *Washington Post*, April 26.

Christakis, Nicholas. 2020. *Apollos's Arrow: The Profound and Enduring Impact of Coronavirus on the Way We Live*. New York: Little, Brown Spark.

Courage, Katherine. 2021. "Covid—Overwhelmed Hospitals Strain Staff and Hope to Avoid Rationing Care." *Scientific American*, January 27.

Cumming-Bruce, Nick. 2021. "Virus Deaths are Probably Two to Three Times More Than Official Records, the W.H.O. Says." *The New York Times*, May 21.

Du, Zhanwei, Xu, Xiaoke, Wu, Ye, Wang, Lin, Cowling, Benjamin, and Meyers, Lauren. 2020. "Serial interval of Covid-19 among publicly reported confirmed cases." *Emerging Infectious Diseases*, 25: 1341–1343.

El-Erian, Mohamed. 2021. "What Can We Learn from the UK's Response to Covid-19?" *The Guardian*, January 25.

El-Khatib, Ziad, Jacobs, Graeme, Ikomey, George, and Neogi, Ujjwal. 2020. "The disproportionate effect of Covid-19 mortality on ethnic minorities: genetics or health inequalities?" *EClinicalMedicine*, 23: 1–2.

Esteve, Albert, Permanyera, Inaki, Boertiena, Diederik, and Vaupel, James. 2020. "National age and coresidence patterns shape Covid-19 vulnerability." *PNAS*, 117(28): 16118–16120.

Frederick, Wayne. 2021. "What Happens When People Stop Going to the Doctor? We're About to Find Out." *The New York Times*, February 22.

Garrett, Laurie. 1994. *The Coming Plague: Newly Emerging Diseases in a World Out of Balance*. New York: Penguin.

George, Judy. 2021. "80% of Covid-19 patients may have lingering symptoms, signs." *Medpage Today*, January 30. https://www.medpagetoday.com/infectiousdisease/covid19/90966

Gettleman, Jeffrey, Yasir, Sameer, Kumar, Hari, and Raj, Suhasini. 2021. As Covid-19 Devastates India, Deaths Go Undercounted." *The New York Times*, April 30.

Gjaja, Marin, Hutchinson, Rich, Farber, Adam, and Brimmer, Amanda. 2020. "Three Paths to the Future," *Boston Consulting Group*, May 27.

Grose, Jessica. 2021. "They're Tired as Hell and They Just Can't Take It Anymore." *The New York Times*, February 7.

Hall, Stephen, Sun, Ying, and Holt, Tania. 2021. "'None Are Safe Until all Are Safe': Covid-19 Vaccine Rollout in Low- and Middle-Income Countries." *McKinsey & Company*, April 23.

Hanage, William. 2022. "After Omicron, This Pandemic Will Be Different." *The New York Times*, January 19.

Holshue, Michelle, DeBolt, Chas, Lindquist, Scott, Lofy, Kathy, Wiesman, John...Pillai, Satish. 2020. "First case of novel coronavirus 2019 in the United States." *The New England Journal of Medicine*, 382: 929–936.

Horton, Richard. 2020. *The Covid-19 Catastrophe*. Cambridge: Policy Press.

Interlandi, Janeen. 2021. "The World Is at War With Covid. Covid Is Winning." *The New York Times*, October 14.

Johnson, Steven. 2021. "The Great Aftermath." *The New York Times Magazine*, November 28.

Klein, Ezra. 2022. "The Covid Policy That Really Mattered Wasn't a Policy." *The New York Times*, February 6.

Leffler, Christopher, Ing, Edsel, Lykins, Joseph, Hogan, Matthew, McKeown, Craig, and Grzybowski, Andrzej. 2020. "Association of country-wide coronavirus mortality with demographics, testing, lockdowns, and public wearing of masks." *American Journal of Tropical Medicine and Hygiene*, 103(6): 2400–2411.

McKenzie, Pete. 2022. "As Cases Skyrocket, New Zealand Finally Faces Its Covid Reckoning." *The New York Times*, March 3.

Mueller, Benjamin. 2021. "As Covid Ravages Poorer Countries, Rich Nations Spring Back to Life." *The New York Times*, May 5.

Partington, Richard. 2021. "UK Unemployment Reaches Four-Year High in Covid-19 Lockdown." *The Guardian*, January 26.

Perry, Brea, Aronson, Brian, and Pescosolido, Bernice. 2021. "Pandemic precarity: Covid-19 is exposing and exacerbating inequalities in the American heartland." *PNAS*, 118(8): e2020685118.

Phillips, Steven and Williams, Michelle. 2021. "Confronting our next national health disaster—long haul Covid." *The New England Journal of Medicine*, 385: 577–579.

Rossen, Lauren, Branum, Amy, Ahmad, Farida, Sutton, Paul, and Anderson, Robert. 2020. "Excess deaths associated with Covid-19, by age and race and ethnicity— United States, January 26—October 3, 2020." *Morbidity and Mortality Weekly Report*, Centers for Disease Control and Prevention, 69(42): 1522–1527.

Serkez, Yaryna. 2021. "These Charts Show Your Lockdown Experience Wasn't Just About Luck." *The New York Times*, March 11.

Sgaier, Sema. 2021. "Meet the Four Kinds of People Holding Us Back from Full Vaccination." *The New York Times*, May 18.

Shaman, Jeffrey. 2022. "What Will Our Covid Future Be Like? Here Are Two Signs to Look Out For." *The New York Times*, March 4.

Sokat, Kezban and Altay, Nezih. 2021. "Serving vulnerable populations under the threat of epidemics and pandemics." *Journal of Humanitarian Logistics and Supply Chain Management*, 11(2): 176–197.

Taub, Amanda. 2020. "Why Are Women-Led Nations Doing Better with Covid-19?" *The New York Times*, May 15.

Tooze, Adam. 2021. *Shutdown: How Covid Shook the World's Economy*. New York: Viking.

Tufekci, Zeynep. 2021. "Why Did It Take So Long to Accept the Facts of Covid?" *The New York Times*, May 7.

Turkewitz, Julie and Taj, Mitra. 2021. "After a Year of Loss, South America Suffers Worst Death Tolls Yet." *The New York Times*, April 29.

Wright, Lawrence. 2021. "The plague year." *The New Yorker*, 11: 20–59.

Yong, Ed. 2020. "Anatomy of an American failure." *The Atlantic*, September, 33–47.

Zimmer, Carl and Ghorayshi, Azeen. 2021. "Studies Suggest Why Omicron Is Less Severe: It Spares the Lungs." *The New York Times*, December 31.

3

ECONOMIC COLLAPSE

Chapter learning objectives

After reading this chapter, you will be able to:

LO1   Explain the effects of economic shutdown.
LO2   Examine the macroeconomic dimensions of economic collapse.
LO3   Evaluate the leveling effects of the coronavirus pandemic.
LO4   Identify the relationship between inequality and the severity of disease.
LO5   Assess the crisis of inequality.
LO6   Consider the lessons of Covid capitalism.

Chapter outline

Shutdown
Macroeconomic dimensions
Leveling effects
Economic inequality and the severity of disease
The crisis of inequality
Lessons of Covid capitalism
Summary

Shutdown

When the coronavirus pandemic hit London, in the early months of 2020, delivery drivers were "driven to destitution," as individuals in lockdown were worried about contact with others (Shead, 2020). The demand for takeout orders plummeted. While demand eventually returned, the lack of orders meant that many drivers, often paid per delivery, were making less than £1 per hour, which

DOI: 10.4324/9781003310075-4

was far below the national minimum wage. When restaurants shut their doors, deliverers sometimes worked 12 hours per day and took home less than £12. At the same time, when interacting with customers, they faced an elevated risk to their physical well-being. They worried about catching the virus. Before public policy provided a means of economic assistance, they participated in the economy but faced the risk of repeated exposure.

The pandemic existed as a shock to not only economies—consumption, production, investment, and technology—but also human health. The crisis exposed individuals and humanity as the anchors of economic and social life, entangling family and work, blurring the boundaries between public and private life. In March 2020, when the novel coronavirus was spreading out of control, many countries implemented an extreme form of policy intervention: **economic shutdown**, which led to the closing of many segments of the economy. Restaurants, bars, gyms, retail establishments, and other businesses either closed or reduced commerce. By closing non-essential forms of business activity, shutdown interventions limited human contact, requiring **non-essential workers** to stop going to their places of employment; however, **essential workers**, employees who were critical in keeping the economy and society functioning, kept reporting. Adam Tooze (2021) argues that SARS-Cov-2, by the standards of historic pandemics, was "not very lethal. What was unprecedented was the reaction."

On a global scale, large parts of public life, the economy, and social activity shut down. In April 2020, the International Labor Organization estimated that 81 percent of the world's workforce experienced some form of restriction (Tooze, 2021). The effects were widespread. As this chapter explains, the pandemic created a quadruple threat: collapse in demand, decrease in supply, workforce disturbance, and a threat to human health.

Modernity

The defining structures of modernity—hospitals, schools, governing institutions, businesses, and consumers—form the basis of the liberal version of both collective order and individual freedom. In addition to institutions and the physical infrastructure, modern conveniences include aircraft, railways, and subways, where we stand in line, scan, surveil, and move as a herd. During the pandemic, the risk of infection placed all of these vessels of modern life in a precarious position. While the virus surged, reopening restaurants, schools, and businesses risked overwhelming hospitals. But shutdown threatened the existing order: "Shutting the doors of the big institutional complexes brought life as we know it to a halt and cast us back on an unfamiliar reliance on small family networks" (Tooze, 2021). While life in virtual spaces flourished, a growing tradeoff between the economy and public health became fraught and contentious.

Because so much economic activity went dormant, a recession began. Businesses reduced hours or announced sweeping closures. Supply-side disruptions and the collapse of demand spread in supply chain networks to factories throughout the world, including Bangladesh, China, India, and Vietnam. Without markets, suppliers, and workers, unsold merchandise accumulated. But for both workers and consumers, shutting down became a matter of rational decision making. Shoppers stopped buying goods in retail outlets and reduced their consumption of services, buying products online and using home delivery.

Noxious contracts

For many workers, especially at the low end of the income spectrum, the pandemic delivered a one-two punch. David Grusky (2021) of Stanford University, writing in *The New York Times*, quoted a 26-year-old worker, who described her economic status during the first infection wave:

> I was working at a gas station, bringing home enough to get me by. And then the Corona hit, my hours got cut, I was only working one or two days, sometimes no days, and then I was out of a job. It literally was hell...I was suffocating in bills.

This first punch entails the loss of a job. But the second punch, a job-safety punch, is unique to a pandemic. After she lost her job at the gas station, she found employment at a care home: "I'm given a group of people that I have to care for. I'll have to pass out breakfast or lunch trays, or they're incontinent, so I'd be giving towels. It's basic care" (Grusky, 2021). This **noxious contract** entails the acceptance of employment with high risk in order to pay bills. The problem with this arrangement is that, during a pandemic, face-to-face work increases the risk of infection. But individuals who avoid exposure require public income assistance, protection against noxious contracts.

Essential workers

The coronavirus pandemic highlighted the reality that economies rely on essential workers. Although many workers may argue that their contributions are essential, some jobs are too important for disruption to occur, even during a pandemic. The list includes workers involved with the safety of human life, the protection of property, and crucial aspects of the economy, including banking, distribution, education, emergency response, energy, agriculture, healthcare, mass transit, pharmacies, and public safety. During the early months of the coronavirus pandemic, praise for essential and frontline workers, such as doctors and nurses, who risked their lives to treat patients, persisted. These individuals left their homes to perform important tasks for others, at great risk to themselves and their families.

With this reliance on essential workers, a tradeoff exists. Even though essential workers keep the economy functioning, many of the individuals, such as food delivery personnel, have low-income status, pre-existing health conditions, and less access to the medical establishment. A pandemic elevates their risk of exposure, while the healthcare system struggles to provide support. This choice keeps essential workers active but compromised, as non-essential workers perform their economic responsibilities from home.

Chapter thesis and organization

During the pandemic, unemployment rose. The working poor were more susceptible to infection and death, compared with individuals who could work from home. These outcomes led to an important reality, which serves as the chapter's thesis: instead of serving as a great leveler, the coronavirus pandemic exacerbated the problem of **inequality**, the unequal distribution of resources and opportunities. At first glance, the argument that a pandemic should reduce inequality is appealing: everyone is susceptible to a novel coronavirus. New pathogens do not discriminate. They are uniform in their devastation. But at second glance, the coronavirus crisis revealed that marginalized members of society experienced greater risk of exposure and death. The reason: inequality exacerbated the economic and health disparities between those who had access to both economic and healthcare resources and those who did not. In fact, according to Amit Kapoor and Chirag Yadav (2020), global pandemics are more likely during times of growing inequality:

> The argument focuses on several trends that have preceded each pestilence in history. These play out in the following manner. Initially, sustained population growth results in population density, which pushes the basic reproduction number of all diseases upwards. At the same time, overpopulation leads to excess supply of labour, which pushes wages downwards. This immiseration has several effects across society. It reduces the nutrition levels of the poor making them less capable of fighting pathogens. The poor also migrate vast distances to cities in search for jobs. The increasing concentration of people in cities becomes a breeding ground for diseases while their movement makes it easier for diseases to travel across regions.

To address the economic outcomes of the coronavirus pandemic, including economic collapse, the chapter discusses macroeconomic dimensions, leveling effects, economic inequality and the severity of disease, the crisis of inequality, and lessons of Covid capitalism.

Macroeconomic dimensions

The previous economic downturn before the coronavirus shutdown, the Great Recession, began in December of 2007 in the United States as a banking and real estate crisis, ravaged housing and financial markets, and spread through other economies. In the United States, it lasted until June of 2009; however, the unemployment rate increased to 10 percent in October of 2009, which was the highest rate since 1982. In the United Kingdom, in November of 2011, the unemployment rate peaked at 8.5 percent. In the United Kingdom, the previous peak, in April of 1993, was 10.6 percent.

Between the Great Recession and the coronavirus pandemic, economies in developed countries experienced strong rates of workforce participation. **Inflation**, a general increase in the price level, remained low. But, from Japan to Europe to North America, the period included policies of monetary easing that reduced the ability of central banks to cut interest rates during times of trouble. Economies remained strong, with rising production, falling unemployment, and economic growth. After the coronavirus pandemic struck, however, macroeconomic dimensions included a decline in economic activity, policy intervention, and economic recovery.

Decline in economic activity

In March of 2020, with the novel coronavirus spreading, governments around the world implemented economic shutdown interventions, closing businesses and reducing economic activity. Global disruptions became apparent. Container ships, bound for Californian, European, and Asian ports, waited for weeks to unload their goods. Factories sat idle. Severed global supply chains reduced automobile production because of a lack of computer chips. Economies could not keep employees at work, shelves full, or services available. A decrease in aggregate supply led to shortages in medical equipment, shipping containers, and lumber. Layoffs, rising unemployment, and falling incomes led to a decrease in aggregate demand. According to Tooze (2021), "If the first-order effect of the pandemic was to reduce our ability to safely supply goods and that put the livelihoods of hundreds of millions of people in jeopardy, the second-order effect came from the demand side." That is, problems on the supply side of the market led to a demand shock, which decreased employment, income, investment, and sales, putting further pressure on the world's economies. The overall impact was disastrous. Consumption, production, and investment declined. But, in the United States, the National Bureau of Economic Research, the arbitrator of U.S. business cycles, declared that the pandemic recession lasted 2 months, which was the shortest recession on record. The downturn occurred during March and April of 2020. But even though the recession was short, it was severe, increasing the unemployment rate in the United States to the highest level since the Great Depression in the 1930s (Casselman, 2021).

Instability in financial markets

When economies contracted, financial markets teetered on the edge of collapse. But central banks pulled economies back from the brink of disaster. They implemented expansionary **monetary policy**, increasing the availability of credit. As a result, financial markets recovered. For the remainder of 2020, financial investors shrugged off news that could derail positive momentum. During 2021, the S&P 500 gained 26.9 percent, the Dow Jones Industrial Average gained 18.7 percent, and the Nasdaq Composite gained 21.4 percent (Jackson and Schmidt, 2022).

Rising unemployment

Around the world, economic shutdowns led to rising unemployment. In the United States, by the end of March 2020, infections surged, 10 million people were out of work, and 6.6 million applied for unemployment benefits. By the end of April 2020, 30 million people were unemployed (Carter, 2021). During 2020, the U.S. unemployment rate peaked at 14.8 percent (Figure 3.1).

Falling employment-to-population ratio

In addition to the unemployment rate, another way to measure the impact of an external shock on the economy is the ratio of employment to population, which demonstrates the share of the population 16 years and older working full- and part-time. In the United States, a dip in the trend in April 2020 demonstrates the impact of the coronavirus pandemic (Figure 3.2).

Global supply chain disruption

Throughout the coronavirus pandemic, with shipping disturbed, warehouses overflowing, prices rising, and trucks without drivers, the great supply chain disruption became apparent. Lazaro Gamio and Peter Goodman (2021) describe the problem: in early 2020, stores and offices closed, production decreased, and

Seasonally adjusted monthly data from January 2005 to December 2020

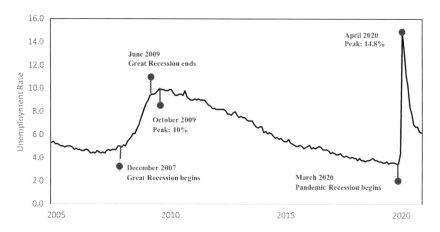

FIGURE 3.1   Historical unemployment rate in the United States.

Source: Author using data from the Federal Reserve Bank of St. Louis, https://fred.stlouisfed.org/

Individuals 16 and older: non-farm employment, seasonally adjusted

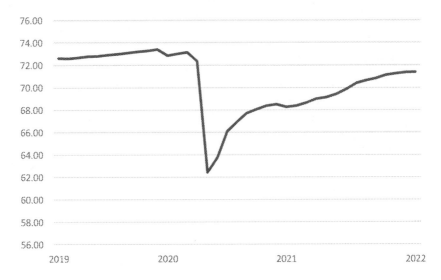

FIGURE 3.2 Employment-to-population ratio in the United States.

Source: Author using data from the Bureau of Labor Statistics, https://data.bls.gov/cgi-bin/
surveymost

layoffs occurred. With a decrease in supply and falling incomes, manufacturers and shipping companies prepared for a decrease in demand. While this occurred, a complicated and nuanced economic reality unfolded. With a surge in infections, hospitals needed surgical masks, gowns, and protective gear. But China manufactures most of these forms of output. As Chinese factories ramped up production, cargo ships delivered the protective equipment to countries around the world, even to areas that have little trade with China. When empty shipping containers accumulated in ports, a shortage of containers emerged in China. At the same time, consumers shifted demand from services to goods. Individuals did not want to eat out or attend events, but they wanted electronic devices and furniture. An increase in demand for durable goods pressured factories in China and other manufacturing centers to produce more output. But when container ships loaded goods and shipped them to ports around the world, they waited in long queues to unload, including ports in Los Angeles and Long Beach. A shortage of both workers at the docks and truck drivers for delivery increased transportation costs. Businesses "struggled to hire workers: at warehouses, at retailers, at construction companies and for other skilled trades. Even as employers resorted to lifting wages, labor shortages persisted, worsening the scarcity of goods" (Gamio and Goodman, 2021). Businesses and consumers then placed orders earlier, straining the system. In retrospect, the supply chain problem revealed weaknesses, requiring several reforms, including "investment, technology and a refashioning of the incentives at play across global business. It will take more ships, additional warehouses and an influx of truck drivers, none of which can be conjured quickly or cheaply" (Goodman, 2022).

Policy intervention

During the coronavirus pandemic, the lessons of the British economist John Maynard Keynes (1883–1946) provided a policy framework. Writing during the Great Depression, Keynes (1936) argued that, during periods of economic contraction, governments should intervene in the economy by stimulating job creation: "For Keynes, the economy was not a self-sustaining engine of prosperity; it was something that societies created to meet social needs and that had to be actively managed to function properly" (Carter, 2021). In Keynes' view, if government did not intervene during economic contraction, a manageable problem could become a national crisis.

Activist government response

Keynes' (1936) prescription created a new **paradigm** in economics, "Keynesianism," which replaced the classical belief that free-market economies would gravitate to full employment. According to the historian of science, Thomas Kuhn (1962), paradigms exist as "universally recognized scientific achievements that for a time provide model problems and solutions to a community of practitioners." Early in the crisis, hospitals faced shortages of personal protective equipment. Problems with test production and distribution plagued efforts to contain the virus. Inequities in vaccine distribution complicated economic recovery. Together, these factors were "not only public health failures but also economic failures—an inability to marshal resources to solve a problem" (Carter, 2021). During the crisis, Keynes' paradigm prevailed. Federal governments, including those in the United States and the United Kingdom, stimulated economic activity, addressed market shortages, and provided financial assistance.

The case of the United States

With increasing government budget deficits, Congress during the Donald J. Trump administration passed in March 2020 a $2.3 trillion relief package and in December 2020 a $900 billion package, examples of **fiscal policy** (government spending). When Joseph R. Biden took the oath of office, on January 20, 2021, his administration established the goal of a return to full employment. The $1.9 trillion American Rescue Plan passed in both houses of Congress. It was signed by President Biden, a Democrat, in March 2021, with zero support from Republicans in Congress.

The American Rescue Plan provided financial aid for businesses and households and accelerated recovery in the labor market. It included aid to state, local, and tribal governments, unemployment benefits, assistance to reopen schools, and resources to fight poverty. An allowance converted a child tax credit into a near-universal benefit. The plan offered subsidies to help low-income households, boosted the earned income tax credits for adults without children, and provided housing vouchers for individuals at risk of homelessness.

This "second war on poverty," initiated during Biden's second month in office, was characterized by his status as a moderate, like President Lyndon B. Johnson, 57 years before him, who oversaw the first war on poverty (Matthews, 2021). In the United States, policy to address poverty was important because one in seven households reported that they did not have enough food (Kristof, 2021). Child poverty cost the country at least $800 billion per year in medical payments and higher crime (National Academies of Sciences, Engineering and Medicine, 2019).

Economic recovery

In 2021, when coronavirus cases fell, vaccines were distributed, and federal aid flowed, consumers found themselves with large levels of savings, resulting from months of lockdown. Over time, the relaxation of shutdown interventions and greater consumer confidence released pent-up demand, including spending on restaurants, hotels, and travel. But the recovery also increased prices of many goods, including chocolate, clothing, diapers, fast food, and gasoline.

The shape of economic recovery

The shape of economic recovery signifies how recoveries are similar to or different from previous patterns. With a V-shaped recovery, the best-case scenario, a rapid expansion follows recession. A U-shape recovery entails a sluggish period of growth. A W-shape means a second recession. An L-shape lacks expansion. After the pandemic recession, the United States experienced a V-shape, measured with a change in real Gross Domestic Product (GDP) (Figure 3.3).

K-shaped recovery

Many economists focus on a k-shaped recovery, distinguishing between those who benefit from recovery (upper portion of the k) and those who do not. Those in the upper portion work from home, experience job stability, and benefit from an appreciation in the value of their assets. The lower portion of the k-shaped recovery represents those who suffer from harmful health outcomes and unemployment (Jones, 2020).

Inflationary pressure

Economic recovery creates a tradeoff between inflation and unemployment: as unemployment decreases, inflation increases. During economic recovery, slow job creation tempers the rise in the general price level; however, full employment (and inflationary pressure) corresponds with rising income:

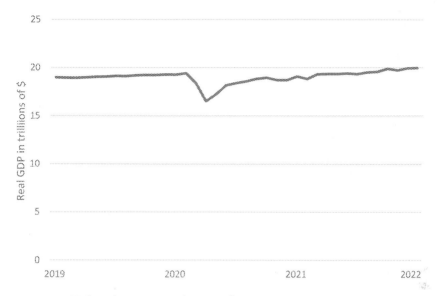

FIGURE 3.3 V-shaped recovery in the United States.

Source: Author using data from YCharts, https://ycharts.com/indicators/us_monthly_real_GDP

> Although it's true that inflation erodes real incomes, there's overwhelming evidence that maintaining full employment is extremely important for reasons that go beyond money. Jobs bring in income; but they also, for many workers, bring dignity, so that being unemployed damages happiness far more than you can explain simply by the lost dollars.
>
> *(Krugman, 2022)*

The problem is that inflation causes hardships for people living paycheck to paycheck. It also erodes the wealth of financial investors. The Consumer Price Index in the United States, a widely used measure of inflation, was 7 percent higher at the end of 2021, almost 2 years into the pandemic, then it was a year earlier, its fastest pace since 1982. Many other countries experienced a rising inflation rate, including Brazil, Canada, and the United Kingdom. While inflation resulted from higher prices for gasoline, hotel rooms, used cars, and many other forms of output, most economic sectors contributed to the rise in the price level. The fear was that inflationary pressure would be difficult to control, becoming entrenched in future expectations.

Even though inflation, as economic analysis reveals, is caused by too much money chasing too few goods, it is a function of both demand-side effects and supply-side effects. During the recovery interval, the question in many economies was whether the price increases bedeviling businesses, consumers, and policymakers resulted more from global factors tied to the pandemic, such as supply chain disruptions, or demand-side policy outcomes.

On the supply side, the supply chain disruptions—including factory shut-downs, shipping delays, and problems of distribution—labor shortages, worker resignations, and retirements decrease aggregate supply: "The conveyor belt that normally delivers goods to consumers suffers from shortages of port capacity, truck drivers, warehouse space and more, and a shortage of silicon chips is crimp-ing production of many goods, especially cars" (Krugman, 2021). Even though global supply chains do not break, if consumers want electronic devices, exer-cise equipment, and materials for home improvement during a pandemic, supply chains cannot keep up. Virus outbreaks disrupt factories, truck driving routes, and ports. As goods in short supply become costlier to transport, prices rise.

In addition to economic bottlenecks, the pandemic entailed both a reluctance of workers to return to their places of employment and early retirements. Paul Krugman (2021), the Nobel-Prize-winning economist, who identifies this trend as "The Great Resignation," identifies the problem as "tight" labor markets, a record numbers of workers quitting because of unacceptable working conditions. In the presence of two factors, workers voluntarily quit: labor shortages increase compensation, and new opportunities exist:

> When workers weigh whether to jump jobs, they don't just assess their own pay, benefits and career development. They look around and take note of how friends feel about the team culture. When one employee leaves, the departure signals to others that it might be time.
>
> *(Goldberg, 2022)*

Peer effects are important. Frustrations may exist over working conditions, a lack of flexibility, and healthcare concerns. A pandemic exacerbates these frustra-tions. With turnover contagion, employers cannot always solve the problem with higher wages. Benefits and flexible working conditions must increase.

On the demand side, in response to the downturn, both fiscal policy of fed-eral governments and monetary policy of central banks stimulate economic activity, sending consumer spending into overdrive. With greater levels of government spend-ing and lower levels of taxation, fiscal policy increases aggregate demand beyond the amount of slack remaining in economies. Monetary policy reduces interest rates, contributing to higher levels of borrowing, especially in residential construction and consumer goods. Low interest rates bolster demand for purchases made on credit, from cars and houses for consumers to computers and equipment for businesses. As a result of fiscal and monetary policy, both individuals and businesses find themselves with more resources, and as they spend the money, an increase in aggregate demand collides with supply chain shortages and tight labor markets. Inflation occurs.

Longer-term outcomes

When individuals move to urban areas and strain the capacity of cities to pro-vide public services, congestion increases exposure to disease. But uncertainties persist. Across human societies, vulnerability varies according to economic development, institutional capacity, and strategies of adaptation. For example, a country may implement changes in the social order that better protect against disease outbreaks. Both the public and private sectors may learn to adapt (Botzen et al., 2019).

In this context of urbanization and vulnerability, the direct effect of a pandemic on economies is the reduction in per-capita output and income when production and employment decline. But this change does not lead to damage to **physical capital**—machines, equipment, and apparatus to produce output—or durable goods. Rather, capital investment during a pandemic may increase production over time. This growth is a function of technological innovation and capital accumulation. As newer and more productive technologies replace older ones, longer-term growth rates rise. But indirect outcomes exist, such as changes in human behavior, institutional arrangements, policy reforms, and positive spillover effects from technological innovation and economic recovery.

When pandemics impact the trajectory of development, a country's behavioral changes influence this pattern: "Only when levels of development have reached a certain point can nations successfully address weak institutions, create better insurance markets, require more stringent building standards, reduce corruption, and institute more advanced warning and emergency response systems" (Kellenberg and Mubarak, 2011). Even though pandemics reduce output and income, disrupt labor markets, and decrease the quality of life, a country's response depends on several factors, including macroeconomic conditions, income inequality, measures of democracy, educational attainment, functionality, fragmentation, and the propensity for collective action. Countries with advanced economies, effective government institutions, and functional social arrangements have the capacity to recover from pandemics.

Leveling effects

To put these trends in perspective, consider that Walter Scheidel (2018) of Stanford University, in *The Great Leveler*, argues that, throughout history, in all societies, the gap between the haves and the have-nots alternatively increases and decreases. From antiquity to the present, in the presence of economic surpluses, those in positions of power do not share the surpluses evenly.

In the past 2 millennia, food production through farming and herding created wealth for landowners. The domestication of animals and plants enabled the accumulation and preservation of productive resources. Social norms and legal systems then made it possible to pass this wealth to future generations.

With these arrangements, several factors impacted family fortunes: consumption, health, investment, marriage, ownership, reproduction, and external shocks. Taken together, these factors, along with effort and luck at the household level and economic systems at the national level, led to unequal long-run outcomes.

For the owners of capital, the processes of globalization, innovation, trade, and urbanization generated positive returns, but not for laborers. Even as economic arrangements, political structures, and social norms changed, inequality remained, or found new ways to persist. In fact, in many examples, such as the United States in the past half century, inequality grew, putting pressure on the public sector to minimize discrimination, underemployment, and social upheaval. The point is that civilization has not created extended periods of peaceful equalization.

But, according to Scheidel (2018), certain historical factors have served as "great levelers," reducing socioeconomic forms of inequality:

> Violent shocks were of paramount importance in disrupting the established order, in compressing the distribution of income and wealth, in narrowing the gap between rich and poor. Throughout recorded history, the most powerful leveling invariably resulted from the most powerful shocks. Four different kinds of violent ruptures have flattened inequality: mass mobilization warfare, transformative revolution, state failure, and lethal pandemics.

Mass mobilization warfare

Violence may level inequality. Small-scale conflicts do not redistribute resources in meaningful ways. But wars mobilize workers and other economic resources. During the twentieth century, the two world wars serve as examples of the mass mobilization of warfare. These conflicts—with large-scale destruction, massive government intervention, confiscatory taxation, disruptions to global trade flows, and inflation—depleted the wealth of elites. The world wars created momentum for public policies that led to equalizing effects, such as the growth of welfare states, progressive taxation, and unionization (Scheidel, 2018).

Transformative revolution

The world wars led to another leveling force, transformative revolutions. In premodern history, peasant revolts typically lacked the force to alter the existing economic order. But, in the modern era, uprisings have sometimes succeeded in calling attention to inequalities. "Violent societal restructuring needs to be exceptionally intense if it is to reconfigure access to material resources" (Scheidel, 2018). In the late eighteenth century, the French Revolution leveled on a relatively small scale. In the twentieth century, the Bolshevik Revolution, accompanied by long-standing campaigns of violence, collectivized and redistributed resources, leveling on a large scale.

State failure

While warfare diminishes the resource base and governing institutions, state failure levels inequality. In many societies, the rich either serve as the ruling class or establish the leadership hierarchy. They are capitalists, allocating resources for personal gain. States establish the legal system and political and economic institutions:

> When states unraveled, these positions, connections, and protections came under pressure or were altogether lost. Although everybody might suffer when states unraveled, the rich simply had more to lose: declining or collapsing elite income and wealth compressed the overall distribution of resources.
>
> *(Scheidel, 2018)*

Lethal pandemics

Lethal pandemics, according to Scheidel (2018), reduce inequality on the largest scale. But bacterial and viral assaults on humanity do not involve violent conflict, revolution, or the failure of the public sector. They are, however, potentially more impactful.

Malthusian framework

In this framework, pandemics reduce inequality by unleashing the positive checks that shorten human lifespans, using the terminology of Thomas Malthus (1798), in *An Essay on the Principle of Population*. In the long run, according to Malthus, population growth outstrips the ability of societies to provide economic resources. This reduces further population growth, involves the preventive checks that decrease fertility, and entails the positive checks that increase mortality. The inventory of positive checks, which includes poverty, disease, famine, and pandemics, contributes to the negative effects on population. Because modern research emphasizes that technological innovation prevents the Malthusian crisis, this framework provides a method to assess the impact of pandemics on premodern societies in late medieval and early modern Europe, 500 AD to 1700 AD.

Black Death

As an example, the Black Death, from the fourteenth century to the seventeenth century, ravaged Europe and parts of Asia, serving as history's best-known pandemic. During the 1320s, plague erupted in the Gobi Desert, in central Asia, caused by a bacterial strain in the digestive tracts of fleas, which infected rodents. Soon after, the rodents transported the plague south to India, east to China, and west to Europe, the Mediterranean, and the Middle East. Estimates of mortality ranged from 25 percent to 45 percent of the human population, tens of millions of people; however, the pandemic did not impact the physical infrastructure. Because production declined less than the population, both per-capita income and output increased. Relative to labor, land became more abundant. While workers benefited at the expense of landowners, the economic shift depended on power structures, markets, and institutions. Decreasing inequality during the time, markets and microbes worked in tandem (Scheidel, 2018).

The coronavirus pandemic as an accelerant

In contrast to the argument that lethal pandemics serve as great levelers, however, the coronavirus pandemic increased inequality. While reducing labor force participation, the crisis did not fundamentally alter the resource base. As a factor of production, labor was still available. Over time, economic recovery lowered unemployment rates to pre-pandemic levels; however, labor shortages persisted. In effect, the pandemic served as an "accelerant," that is, it accelerated "dynamics already present in society" (Galloway, 2020). The dynamic already present included income and wealth inequality and the resulting ill effects.

During the coronavirus pandemic, the susceptibility of marginalized groups to higher levels of infection and death was a function of economic and social inequality. Members of lower-income classes were more likely to serve as frontline workers, live and work in closer contact with others, experience higher levels of economic instability, and suffer from higher rates of exposure. For these individuals, the economic, health, and social problems during the pandemic resulted from pre-pandemic disparities of income, opportunity, and wealth. Because of deprivation, discrimination, and segregation, members of marginalized communities were more likely to suffer from higher levels of morbidity and mortality.

Economic inequality and the severity of disease

In a modern pandemic, once interventions and vaccines slow the spread of disease, the poor are still likely to suffer. Two reasons exist. First, according to research by Davide Furceri of the International Monetary Fund and his colleagues (2020), a global pathogen leaves economic systems more unequal: "major epidemics in this century have raised income inequality, lowered the shares of income going to the bottom deciles, and lowered the employment-to-population ratio for those with basic education but not for those with advanced degrees." This argument is counter to the "great leveler" thesis for pandemics of Walter Scheidel (2018). The empirical evidence that pandemics increase inequality relates to the Gini coefficient, a measure of inequality of a system. Furceri et al. (2020) find that, over time, pandemics lead to persistent increases in this measure: shares of income flowing to the bottom quintile decrease, while the shares flowing to the top quintile increase.

Second, according to Amit Kapoor and Chirag Yadav (2020), because of losses in education, healthcare, and income during a pandemic, poor households struggle. In educational systems, the poor have less access to electronic devices and Internet connections, which are crucial factors for the delivery of online content. Higher levels of pre-existing conditions, less preventive care, and inadequate medical oversight decrease the ability of the poor to receive adequate healthcare. During a pandemic, reductions in employment and income disproportionately impact the most vulnerable members of society. As a result, "the losses from pandemics are more permanent for the poor. These effects widen and cement the gap across various income levels" (Kapoor and Yadav, 2020).

In the research by Furceri et al. (2020) and Kapoor and Yadav (2020), privileged individuals benefit from the existing order, but members of marginalized communities do not. A pandemic widens pre-existing forms of inequality. But this process is not a systematic flaw. Rather, it exists as an outcome in countries that do not have public sectors that address these shortcomings through progressive systems of redistribution.

Costs to public health

The link between inequality and the severity of disease relates to socioeconomic determinates, including education, employment, income, race/ethnicity, sex, and social class. Individuals with higher risk factors experience marginalized status relating to these factors. But marginalized status correlates with respiratory illnesses, including Covid-19. The reasons include the prevalence of chronic respiratory disorders, housing complexes with communal facilities, a lack of insurance, less access to healthcare institutions, poor physical health, and higher levels of poverty. When a local disease outbreak becomes a national epidemic and then a global pandemic, the consequences of these economic and social imbalances become more acute (Nassif-Pires et al., 2020).

Costs to economic well-being

Low-income households struggle to address the problems that emerge during a pandemic, especially volatile employment and falling incomes. These problems result from a lower capacity to mitigate the effects of the pandemic. Because of the availability of fewer resources, it is difficult for the working poor to stock up on food and other necessities, minimize the number of trips to grocery stores, and limit exposure to a spreading pathogen (Nassif-Pires et al., 2020).

Deprivation

The Organization for Economic Cooperation and Development (OECD) Family Database highlights the reality that, in the third decade of this century, more than 13 percent of children across OECD countries live in relative poverty, mostly in households with single parents. When a crisis occurs, these individuals experience a disproportionate burden. Relative deprivation entails less access to running water, sanitation, and clean air. For individuals experiencing material deprivation, basic hygiene may not exist. When communities implement shutdown protocols, poorer members of society struggle to comply. Their jobs require face-to-face interaction. Their wages flow to the consumption of necessities. They may not have a social safety net. Even vaccines favor individuals connected to formal healthcare networks.

Case study 3.1 Deprivation and Covid-19 in the United Kingdom

In the United Kingdom, the first case of Covid-19 was reported on January 31, 2020. On March 23, the national government implemented a lockdown, followed by a series of interventions, including school closings, mask mandates, and a tiered system of alerts. Over time, the coronavirus crisis highlighted pre-existing health inequalities, especially in Black and Asian communities, where relatively higher levels of Covid-19 deaths and infections related to the national public health response and socioeconomic status of individuals. Research on the impact of Covid-19 found that people in more disadvantaged areas were less likely to comply with shutdown interventions, due to either their occupation or mistrust of authorities (Morrissey et al., 2021). In this context, risk factors leading to Covid-19 morbidity and mortality existed at the intersection of individual and community levels, involving both economic and social factors, including less access to healthcare and education, occupation, poor housing, and unemployment. The Office for National Statistics examined deaths from Covid-19, according to geographies and levels of deprivation, a measure of relative poverty. During the first half of 2020, at the beginning of the pandemic, the mortality rate for individuals with Covid-19 in the most deprived areas was more than double the rate for individuals in the least deprived areas: 55 deaths per 100,000 residents versus 25 deaths per 100,000 residents (Horton, 2020). During this time, London had the highest age-standardized mortality rate of deaths from Covid-19 in the United Kingdom and urban areas had a higher mortality rate than rural areas. Varying levels of vulnerability at the community level impacted pandemic responses and recovery. Because relative deprivation correlated with inferior health outcomes, it led to a disproportionate burden on low-income members of society.

A crisis of inequality

In the United States, despite a record-low unemployment rate of 3.5 percent in February 2020, before the pandemic, Joseph Stiglitz, the Nobel-Prize-winning economist at Columbia University, argued that "Years of limp wage growth left (many) workers struggling to afford essentials. Irregular work schedules caused weekly paychecks to surge and dip unpredictably. Job-based benefits were threadbare or nonexistent" (Cohen, 2020). In this and other countries, globalization, deindustrialization, and changing labor market conditions created a crisis of inequality.

The rich get richer

During the economic shutdown interval of the coronavirus pandemic, millions of people lost their jobs. Restaurants, bars, and clubs closed, retail establishments offered online buying options, and manufacturing plants reduced production. Yet an unprecedented trend appeared: despite the global nature of the economic downturn, the rich got richer. One of the most important economic characteristics of the coronavirus pandemic was a greater concentration of income flowing to the richest households.

Knowledge workers, individuals who use information in the workplace, have higher incomes than average. During the coronavirus pandemic, they were able to work remotely. During lockdown, many individuals working for tech and other large companies adjusted their schedules to avoid the office. These well-positioned employees benefited from new habits developed during the pandemic with remote work. Households ordered items for delivery. Streaming shows provided entertainment options. Teleconferencing and file sharing offered methods to enhance work. According to the *Wall Street Journal* (2021), the deep pockets of tech and other large companies enabled them to thrive during the first year of the coronavirus pandemic, developing products and services that appealed to consumers in an economy in flux:

> The result was dizzying growth for some of the largest corporations in history—and for their stock prices. At a time when companies such as airlines and bricks-and-mortar retailers struggled to survive, combined revenue for the five biggest U.S. tech companies—Apple Inc., Microsoft Corp., Amazon.com Inc., Google-parent Alphabet Inc., and Facebook Inc.—grew by a fifth, to $1.1 trillion. Their aggregate profit rose an even faster 24%. And their combined market capitalization soared by half over the past year to a staggering $8 trillion.

Insufficient resources

For the working poor, economic resources are often insufficient to meet daily needs. Service-sector jobs, such as food delivery and cooking, offer low pay and no benefits. But when it comes to the working poor, a willingness to work is not the problem. On the contrary, for the working poor, it is difficult to share in the process of economic growth. David Leonhardt and Yaryna Serkez (2020) argue that, between 1980 and 2020, the U.S. economy almost doubled in size, as measured by GDP per capita, adjusted for inflation and population growth. But this measure conceals an unequal distribution of economic gains.

Disparities in income

In the United States, between 1980 and 2020, the after-tax income rose by 20 percent for the bottom half of the income distribution but 420 percent for the top

0.01 percent (Leonhardt and Serkez, 2020). For individuals in the lowest income quintile, a reason for this disparity is the polarized nature of job opportunities. For both high-skill, high-wage occupations (consulting, finance, technology) and low-skill, low-wage occupations (construction, delivery, personal services), an increase in the demand for labor is occurring. But the demand for labor is declining for many middle-wage, middle-skill opportunities (clerical, retail, sales). The implication is a decrease in the economic position of workers without a university education, who do not possess a trade skill or participate in growing markets (Autor, 2010).

Disparities in wealth

Since 1990, the richest 10 percent of households in the United States, who own more than 80 percent of U.S. stocks, tripled their wealth. At the same time, the bottom 50 percent of households, which rely on wages and salaries, experienced zero gains in their wealth (Friedman, 2021). In 2016, the median American household had a 30 percent lower net worth than in 2007 (Leonhardt and Serkez, 2020). Given a growing economy and rising stock market, how was this possible? The most affluent members of society benefited the most from financial-market gains: the richest 0.1 percent of households possess 20 percent of the country's wealth, up from 7.4 percent in 1980. Joseph Stiglitz (2013), in *The Price of Inequality*, argues that the trend will continue.

Disparities in socioeconomic conditions

Trends in health care, housing, and life expectancy demonstrate inequalities. For lower-income households, in the United States, healthcare has increased as a percentage of total spending. During the first 2 decades of this century, the number of housing units renting for more than $1,000 increased by 5 million. In the United States, 10 million families struggle to afford an apartment. When rent is more than 50 percent of their take-home pay, these families scrimp on healthcare, transportation, and food. Americans in the bottom fourth of the income distribution live, on average, 13 years less than those in the top fourth (Leonhardt and Serkez, 2020).

Lessons of Covid capitalism

The coronavirus pandemic increased morbidity and mortality but impacted a smaller share of the world's population than previous pandemics such as the Justinian Plague of the sixth century that killed one-half of the world's population, Black Death of the fourteenth century that killed one-fourth of the world's population, or the influenza pandemic of the early twentieth century that killed 50 million people. Creating negative health outcomes, the coronavirus crisis reduced the production of output. An increase in labor scarcity put upward

TABLE 3.1 Economic lessons

Topic	Economic lesson
Collective action	Health and safety depend on collective action
Density caps	Density caps reduce infections and maintain business activity
Employment modifications	Pandemics create flexible working conditions
Information gap	Pandemics exist as problems of information
Intersectionality	Intersectional pressures complicate pandemic outcomes
Reimagining capitalism	Society should reimage the system of capitalism
Remote work	A pandemic creates context for remote work
Working women	Economic and social policy should support working women

Source: Author.

pressure on wages but did not alter the prospects of the working poor. The most important lesson, however, extended beyond public health and economic collapse: the pandemic accelerated transitions already underway. In particular, innovation and unrest created space for development and change. Over time, new companies, markets, and production methods emerged. In this context, lessons of Covid capitalism relate to collective action, density caps, employment modifications, the information gap, intersectionality, capitalism, remote work, and working women (Table 3.1).

Health and safety depend on collective action

Reflecting on the public health disaster during the coronavirus pandemic, Jeneen Interlandi (2020) argued for collective action. As the crisis unfolded, many countries, including the United States and the United Kingdom, struggled to minimize damage effects. Shortages of medical equipment risked the lives of doctors and nurses. Shortages of intensive care beds made it impossible to treat all patients. Uncoordinated responses between federal and state governments limited testing capacity. An inability of political leaders to provide a coherent message muddled the importance of masks, quarantines, and social distancing. But a successful response required coordination. Government, the private sector, and households needed to work in tandem. Local health departments, including epidemiologists, health technicians, and public information specialists, required resources. Health networks needed new technologies, innovative methods, web portals for information dissemination, rapid diagnostic tests, and equitable processes of deployment. These problems highlighted the fact that, with future crises, marginalized communities should not experience a disproportionate burden. "But if Covid-19 has taught us anything, it's that our health and safety depend on collective action. That's what public health is all about" (Interlandi, 2020).

Density caps reduce infections and maintain business activity

To fight the global spread of disease, public sectors implement household lock-downs and economic shutdowns. But research from scientists at Northwestern University and Stanford University demonstrate that density caps—limits on the capacity of establishments—reduce infections while maintaining business activity (Chang et al., 2021). These occupancy limits do not stop the spread of virus transmissions. They work in some contexts better than others. However, they bolster struggling economies and reduce the spread of infections. In addition to less access to healthcare resources, more pre-existing conditions, and inferior socioeconomic status, the mobility patterns of low-income households cause disparities in pandemic outcomes: "Density caps (do not) eliminate inequality. But they counteract its effect on the pandemic, preventing dangerous high density interactions that drive disease spread among lower-income populations" (Serkez, 2020).

Pandemics create flexible working conditions

During the coronavirus pandemic, millions of jobs were modified. This reality established a large-scale need for flexible working conditions. As an example, in the professional class, the crisis triggered a permanent shift in where and how people work. The private sector moved forward with plans for employees to commute less, work more from home, reduce business travel, and establish flexible schedules. This shift altered the role of supporting economies in business districts, including hotels, offices, and restaurants. But, as workers retrained and/or changed careers, automation accelerated, and fewer jobs existed in segments of the economy that once provided stable employment opportunities, including food services, human resources, retail sales, and secretarial staff. At the same time, as online commerce boomed, companies increased programmers and search engine optimizers, and warehouses added delivery jobs. In this economy, spending on durable goods increased, including cars and trucks, cell phones, home furnishings, housing, and laptops.

> In every recession, shifts take place in the composition of economic activity; the economy rarely looks the same after a wrenching event as it did before. But what is striking (about the coronavirus pandemic) is the scale and speed of the economy's rewiring.
>
> *(Irwin, 2021)*

Pandemics exist as problems of information

Joshua Gans (2020) of the University of Toronto argues that "pandemics are information problems" with two challenges: identifying who is initially spreading the virus and establishing appropriate responses. With SARS-Cov-2, the average latency period (the time between initial infection and when an individual can spread the virus) was shorter than the average incubation period (the time between infection and when an individual shows symptoms). Even though individuals who were exposed did not initially show symptoms, they could infect others. Before public sectors intervened, a rise in morbidity and mortality overwhelmed hospitals, created medical equipment shortages, and forced economic shutdowns. In this context, an information gap, when society is missing information necessary to make an informed response, creates the conditions necessary for a lethal pathogen to spread through a population. But once it is clear that policy interventions address the problem, society may struggle to implement an optimal policy response. Even in countries with advanced economies such as the United Kingdom and the United States, nuanced approaches were rare. Important tradeoffs occurred. Restrictions on economic activity increased unemployment and anxiety. Lockdowns increased isolation and domestic violence. A focus on treating the victims of the pandemic reduced the number of routine checkups. The result was an ill-informed decision-making apparatus that struggled to balance public health and economic interests.

Intersectional pressures complicate pandemic outcomes

As Chapter 1 explains, intersectionality—the interconnected nature of social categorizations—means some individuals have multiple forms of marginalized status. First, as economies improved, businesses reopened, and consumers resumed consumption activity, more than a year after the pandemic began, women with minority status, such as Black and Hispanic women, were often working less compared with their white counterparts. Second, during the early months of the pandemic, individuals with bachelor's degrees were less likely to lose their jobs. Industries that transitioned to remote work, such as higher education and technology, favored those with more education. Third, many people who lost their jobs earned relatively lower wages. Workers in the lowest income quintile were much more likely to suffer from unemployment. Overall, these problems "put pressure on the cracks in (the) economy that already existed" (Koeze, 2021).

Society should reimagine the system of capitalism

By reimagining the system of capitalism, Mariana Mazzucato (2013), an economist at University College London, influences business leaders and policy makers. She argues in her book, *The Entrepreneurial State: Debunking Public vs Private Sector Myths*, that (1) it is important to define economic growth as a broad measure and (2) many of the world's greatest achievements, such as the invention of the Internet, require contributions from the public sector. The coronavirus pandemic highlights the need to consider Mazzucato's argument. In particular, the crisis reveals weaknesses in the global economy, including fragile supply chain networks, a problem of collaboration across economies, and a reliance on unpaid labor. But maintaining a balance between innovation-fueled economic activity and strong funding, oversight, and research from public sectors, according to Mazzucato, means economies would increase the value of

> Our most valuable, irreplaceable citizens . . . those who work in health and social care, education, public transport, supermarkets and delivery services. These jobs are disproportionately occupied by women, as well as by people of color, in Europe, the U.K. and the U.S.
>
> *(Gupta, 2020)*

A pandemic creates context for remote work

Shutdowns encourage remote work. But this change creates both costs and benefits. On the benefit side, workers in their homes maintain or enhance their levels of **productivity**, their output per unit of time. The elimination of commutes and inefficiencies in the marketplace lead to welfare gains. With a reduction in office space, companies require less capital to generate the same level of output. On the cost side, young employees do not receive the face-to-face mentoring necessary to learn about the work environment. Impersonal digital interaction does not replace face-to-face contact. External effects impact related industries. When fewer workers commute, eat out for lunch, and shop in business districts, market activity such as restaurant meals, taxi rides, and retail shopping decline. As a result, city centers suffer. Given the costs, however, the benefits of remote work demonstrate the importance of the practice.

Economic and social policy should support working women

Despite policy efforts to control the coronavirus, it was difficult to disentangle one crisis from another, a recurring theme in this book. Kim Brooks (2020) argues that "Pandemics make visible what's been hidden; they illuminate the connections between us." She emphasizes the "costs, contradictions and compromises" that working women face, when balancing careers and family. Beyond the challenge of establishing a sustainable work-life balance, the coronavirus pandemic exposed, according to Brooks, the weaknesses of economic systems. Most professional gender gaps are motherhood gaps. In the United States, women earn 82 cents for every dollar that men make, but women without children are closer to parity. The Harvard economist Claudia Goldin (2014) finds that women in their thirties, among the prime childbearing years, experience the largest wage gap. Workplaces penalize women who require flexible hours. During the first year of the coronavirus pandemic, 2.5 million women left the U.S. labor force, compared with 1.8 million men (Rogers, 2021). Several reasons existed. First, more women worked in industries such as hospitality, which were less suited to physical distancing. Second, women were more likely to assume responsibility for children. Third, early childhood education programs were closed during the shutdown interval. As a result, if working mothers were not laid off, they had to perform many jobs at once, such as childcare, housework, and paid employment, which increased anxiety, frustration, and economic inequality (Taub, 2020).

Summary

The coronavirus pandemic led to a period of economic collapse. A decline in both aggregate demand and aggregate supply decreased the production of output, reduced employment, and altered the structure of economies. The crisis accelerated dynamics already underway. But the outcomes were uneven. Frontline workers who experienced a greater risk of exposure to the virus had to continue to report to work to receive a paycheck and scramble to oversee education for their children. These individuals were in inferior economic positions. But members of the professional class who worked from home benefited from work flexibility and the ownership of financial assets. Overall, the coronavirus pandemic did not serve as a great leveler. Rather, it accelerated pre-existing forms of inequality. Collective action, density caps, modifications in employment, information gaps, intersectionality, remote work, reimagining capitalism, and supporting working women provide a set of lessons of Covid capitalism.

Chapter takeaways

LO1   Economic shutdown closes many segments of the economy.

LO2   The macroeconomic dimensions of the coronavirus pandemic include a decline in economic activity, policy interventions, and economic recovery.

LO3   In history, certain factors serve as great levelers, including lethal pandemics, reducing socioeconomic forms of inequality; however, the coronavirus pandemic exists as an accelerant of inequality.

LO4   In the contemporary environment, a global pandemic creates more inequality.

LO5   Globalization and deindustrialization create a crisis of inequality.

LO6   Lessons of Covid capitalism relate to collective action, density caps, employment modifications, the information gap, intersectionality, reimagining capitalism, remote work, and the support of working women.

Key terms

Economic shutdown	Non-essential workers
Essential workers	Noxious contract
Fiscal policy	Paradigm
Inequality	Physical capital
Inflation	Productivity
Monetary policy	

Questions

1   For the working poor, list and characterize the economic and social pressures that existed during the coronavirus pandemic. How did the lives of these individuals differ from members of higher socioeconomic classes?
2   In what sense do pandemics exist as problems of information?
3   How do pandemic recessions impact the economy? Do the costs of slowing the spread of the virus exceed the benefits? Why or why not?
4   With respect to income and wealth inequality, characterize current trends.
5   What factors serve as "great levelers"? Why? Is the coronavirus pandemic an example of a great leveler? Explain.
6   How did the coronavirus pandemic and corresponding economic collapse impact public health and economic well-being?
7   From an intersectional perspective, how do pandemic recessions impact the most vulnerable members of society?
8   Lessons of Covid capitalism include several topics. Explain these lessons. Do others exist?

References

Autor, David. 2010. "The Polarization of Job Opportunities in the U.S. Labor Market: Implications for Employment and Earnings." *The Hamilton Project*. Center for American Progress.

Botzen, Wouter, Deschenes, Olivier, and Sanders, Mark. 2019. "The economic impact of natural disasters: a review of models and empirical studies." *Review of Environmental Economics and Policy*, 13(2): 167–188.

Brooks, Kim. 2020. "Feminism Has Failed Women." *The New York Times*, December 23.

Carter, Zachary. 2021. "We Found the Money We Needed." *The New York Times*, March 14.

Casselman, Ben. 2021. "Officially, the Pandemic Recession Lasted Only Two Months." *The New York Times*, July 21.

Chang Serina, Pierson, Emma, Koh, Pang, Gerardine, Jaline, Redbird, Beth, Grusky, David and Leskovev, Jure. 2021. "Mobility network models of Covid-19 explain inequities and inform reopening." *Nature*, 589: 82–87.

Cohen, Patricia. 2020. "Straggling in a Good Economy, and Now Struggling in a Crisis." *The New York Times*, April 16.

Friedman, Thomas. 2021. "Made in the U.S.A.: Socialism for the Rich. Capitalism for the Rest." *The New York Times*, January 26.

Furceri, Davide, Loungani, Prakash, Ostry, Jonathan, and Pizzuto, Pietro. 2020. "Covid-19 Will Raise Inequality if Past Pandemics Are a Guide." *VoxEU*, May 8.

Galloway, Scott. 2020. *Post Corona: From Crisis to Opportunity*. New York: Portfolio/Penguin.

Gamio, Lazaro and Goodman, Peter. 2021. "How the Supply Chain Crisis Unfolded." *The New York Times*, December 5.

Gans, Joshua. 2020. *The Pandemic Information Gap*. Cambridge, MA: The MIT Press.

Goldberg, Emma. 2022. "We're Out of Here!" *The New York Times*, January 23.

Goldin, Claudia. 2014. "A grand gender convergence: its last chapter." *American Economic Review*, 104(4): 1091–1119.

Goodman, Peter. 2022. "A Normal Supply Chain? It's 'Unlikely' in 2022." *The New York Times*, February 1.

Grusky, David. 2021. "The Pandemic Economy and the Rise of the 'Noxious Contract.'" *The New York Times*, March 9.

Gupta, Alisha. 2020. "An 'Electrifying' Economist's Guide to the Recovery." *The New York Times*, November 19.

Horton, Richard. 2020. *The Covid-19 Catastrophe*. Cambridge: Policy Press.

Interlandi, Jeneen. 2020. "The U.S. Approach to Public Health: Neglect, Panic, Repeat." *The New York Times*, April 9.

Irwin, Neil. 2021. "The Economy Is (Almost) Back. It Will Look Different Than It Used To." *The New York Times*, April 29.

Jackson, Anna-Louise and Schmidt, John. 2022. "2021 Stock Market Year in Review." *Forbes Advisor*, January 3.

Jones, Chuck. 2020. "Three Charts Show a K-Shaped Recovery." *Forbes*, October 24.

Kapoor, Amit and Yadav, Chirag. 2020. "View: Inequality and Pandemics." *The Economic Times*, July 8.

Kellenberg, Derek and Mubarak, Mushfiq. 2011. "The economics of natural disasters." *Annual Review of Resource Economics*, 3: 297–312.

Keynes, John M. 1936. *The General Theory of Employment, Interest, and Money*. London: Macmillan.

Koeze, Ella. 2021. "A Year Later, Who Is Back to Work and Who Is Not?" *The New York Times*, March 9.

Kristof, Nicholas. 2021. "When Biden Becomes…Rooseveltian!" *The New York Times*, January 17.

Krugman, Paul. 2022. "The Secret Triumph of Economic Policy." *The New York Times*, January 13.

Krugman, Paul. 2021. "The Year of Inflation Infamy." *The New York Times*, December 19.

Kuhn, Thomas S. 1962. *The Structure of Scientific Revolutions*. Chicago, IL: The University of Chicago Press.

Leonhardt, David and Serkez, Yaryna. 2020. "America Will Struggle after Coronavirus. These Charts Show Why." *The New York Times*, April 10.

Malthus, Thomas. 1798. *An Essay on the Principle of Population*. Cambridge: Cambridge University Press.

Matthews, Dylan. 2021. "Joe Biden Just Launched the Second War on Poverty." *Vox*, March 10.

Mazzucato, Mariana. 2013. *The Entrepreneurial State: Debunking Public vs Private Sector Myths*. London: Anthem Press.

Morrissey, Karyn, Spooner, Fiona, Salter, James, and Shaddick, Gavin. 2021. "Area level deprivation and monthly Covid-19 cases: the impact of government policy in England." *Social Science & Medicine*, 289: 114413.

Nassif-Pires, Luiza, De Lima Xavier, Laura, Masterson, Thomas, Nikiforos, Michalis, and Rios-Avila, Fernando. 2020. *Pandemic of Inequality*. Public Policy Brief 149. Annondale-on-Hudson: Levy Economics Institute of Bard College.

National Academies of Sciences, Engineering, and Medicine. 2019. *A Roadmap to Reducing Child Poverty*. Washington, DC: The National Academies Press.

Rogers, Katie. 2021. "2.5 Million Women Left the Work Force During the Pandemic. Harris Sees a 'National Emergency.'" *The New York Times*, February 18.

Scheidel, Walter. 2018. *The Great Leveler: Violence and the History of Inequality from the Stone Age to the Twenty-Fist Century*. Princeton, NJ: Princeton University Press.

Serkez, Yaryna. 2020. "A 20 Percent Shutdown Could Be Enough." *The New York Times*, December 17.

Shead, Sam. 2020. "Driven to Destitution: Delivery Riders in Britain are Struggling as Takeout Orders Plummet." *CNBC*, April 23.

Stiglitz, Joseph. 2013. *The Price of Inequality*. New York: W. W. Norton & Company, Inc.

Taub, Amanda. 2020. "Pandemic Will 'Take Our Women 10 Years Back' in the Workplace." *The New York Times*, September 26.

Tooze, Adam. 2021. *Shutdown: How Covid Shook the World's Economy*. New York: Viking.

4

CLIMATE CATASTROPHE

Chapter learning objectives

After reading this chapter, you will be able to:

LO1   Link the coronavirus pandemic to climate change.
LO2   Explain the Hothouse Earth scenario.
LO3   Discuss the economics and science of climate change.
LO4   Address climate effects, including droughts, extreme weather events, rising sea levels, rising temperatures, and wildfires.
LO5   Describe the economic and social impacts of climate change.
LO6   Characterize the quadruple squeeze.
LO7   Express different strategies of resilience.

Chapter outline

The coronavirus pandemic and climate effects
Hothouse Earth
Economics and science of climate change
Climate effects
Economic and social impacts
Quadruple squeeze
Resilience
Summary

The coronavirus pandemic and climate effects

During the coronavirus pandemic, wildfires ravaged the landscape in northern Nevada. But the intermountain valley region of the state geographically

DOI: 10.4324/9781003310075-5

restricted the dispersion of pollutants from the conflagrations. What was the result? "(M)any residents had prolonged exposure to smoke containing elevated levels of particulate matter" (Kiser et al., 2021). This exposure, however, increased human susceptibility to respiratory viruses, including SARS-Cov-2, via modified immune responses. In effect, "Wildfire smoke . . . greatly increased the number of Covid-19 cases" (Kiser et al., 2021). During this time, both the coronavirus pandemic and wildfires bedeviled residents of northern Nevada.

As this chapter explains, the period of time during the coronavirus pandemic produced damaging weather events of unusual and unprecedented ferocity, including wildfires, extended droughts, and heat waves. While the aforementioned story of wildfires and human health provides an example of interconnection, **climate change**—the long-term shifts in temperature and weather patterns—exists as the world's most important environmental problem.

In the contemporary environment, rising average global temperatures lead to recurring droughts, fires, floods, and human misery. The failure of humanity's response is clear: "Despite repeated warnings going back decades, we are not addressing the greatest challenge the planet faces with anything approaching the response it requires. Climate change is already here; it's just not evenly distributed yet. Nor will it ever be" (Editorial, 2022). Many of the countries that are most vulnerable to the effects of climate change, especially those with a developing-economy status, possess the least control over rising temperatures, because they emit few **greenhouse gas emissions**, the heat-trapping gases that warm the planet. It is the responsibility of the United States, China, the European Union, and other developed countries to respond to the problem.

Fossil fuel era

During the fossil fuel era—with its foundations in the nineteenth century's oil industry—greenhouse gas emissions result from fossil fuel combustion. The result of a rise in the atmospheric concentration of greenhouse gases is an increase in average global temperature. Many pathways exist for fossil fuels, including coal, oil, and natural gas, to flow into the global economy. For example, the world consumes 100 million barrels of oil per day, mostly in transportation and power generation. But those who argue that we must alter the trajectory of the climate emergency often make the case for **decarbonization**, severing the link between fossil fuels and economic activity. In transportation and power generation, the most important way to accomplish this goal is to increase the demand for electric vehicles and renewable forms of energy, such as solar and wind power.

Through decarbonization and energy-sector transformation, the idea is to end the fossil fuel era. But while energy sectors increase the capacity of solar and wind, the prices of renewables fall, and the demand for electric vehicles grows, the global economy still relies on fossil fuels. As a result, breaking the chain of transmission between the carbon economy and the climate exists as a long-term problem. With climate change, relaxing vigilance invigorates the factor that causes the problem: carbon emissions from the consumption of fossil fuels. With the coronavirus pandemic, relaxing vigilance in the fight against the virus prolonged the crisis.

Climate change contributes to pandemics

With climate change, pandemics may become more common. Most emerging infectious diseases originate in animals. With deforestation, the expansion of agricultural land, hunting wild animals, and intensification of livestock production, animal-borne diseases spread. It is, therefore, important to address the link between changes in the environment and the onset of infectious diseases.

As the planet warms, viruses such as SARS-Cov-2 spread in places that harbor the species that give rise to the viruses. For example, climatic shifts in southern China—where SARS-Cov-2 originated—enhance bat biodiversity, increasing the number of bat-related coronaviruses that jump to the human population. In southern China, the changing climate alters patterns of vegetation, the distribution of species, and temperature, all factors that enhance bat habitats.

As these habitats reconfigure, bats leave their areas of residence, carry their pathogens, and establish new methods of interaction with humans. As a result of greater human encroachment into natural environments, the pathway between humans and bats establishes a mechanism in which climate change enhances the potential for disease outbreaks. The key is that climate change contributes to the factors that bring pathogens closer to the human population (Beyer et al., 2021).

Pandemics alter the process of climate change

Pandemics also alter the process of climate change. First, economic shutdown reduces fossil fuel consumption. Greenhouse gas emissions decline. During the lockdown phase, individuals drive less. In China, during February 2020, at the beginning of the pandemic, a decrease in driving led to a 25 percent decline in carbon emissions, equivalent to 200 million tons of **carbon dioxide** (CO_2) and equal to more than half of the annual emissions of the United Kingdom. Second, shutdown measures decrease electricity consumption. During economic shutdown, in many countries, peak rates declined. Third, with a decline in the demand for transportation, air quality improves. Cities experience cleaner skies. But, when economies recover, they return to pre-existing patterns. Both fossil fuel consumption and greenhouse gas emissions increase. During the oil shocks of the 1970s and the Great Recession of 2008, similar dips in economic activity and greenhouse gas emissions occurred. But after these periods, previous patterns of economic and environmental behavior resumed.

Intersections

By creating disproportionate impacts on the most vulnerable members of society, including the elderly, homeless, incarcerated, poor, sick, stateless, and unemployed, pandemics and climate change intersect. For example, to fight wildfires, the state of California uses prison inmates to supplement its firefighting force. With 3-foot chain saws and 60-pound packs, the inmates charge into fire zones. During the coronavirus pandemic, however, as climate change increased the frequency and intensity of wildfires, many inmates went home in early release programs. The idea was to protect inmates from the coronavirus, which was spreading through prisons. But a side effect was a decrease in the ability of the state to control wildfires. In addition, during the pandemic, in the San Joaquin Valley, the fertile area in Central California that serves as the nation's breadbasket, when wildfires were burning, rising heat made working conditions unbearable. Smoke settled into the air. Pickers started at 4 am in the fields and worked throughout the day. During this time, wildfires impacted some of the poorest and most neglected laborers. During 2020, when more than 7,000 fires in the state scorched 1.4 million acres, the novel coronavirus ravaged immigrant communities, including pickers in the San Joaquin Valley. In this area, summer days are hotter than they were a century ago. The nights, when individuals normally cool down, are also hotter. Wildfires compound these problems, complicating the working conditions of the most vulnerable laborers.

The world's most important environmental problem

Climate change serves as the world's most important environmental problem. Droughts, rising sea levels, heat waves, and wildfires occur on a more frequent basis. Droughts, for example, increase food insecurity, human migration, and the threat of violence. Rising sea levels flood island countries and coastal cities, displacing communities, inundating thousands of acres of land, and creating billions of dollars in losses. Major world cities, including Amsterdam, Hong Kong, Melbourne, Miami, New Orleans, New York City, and Tokyo, experience the threat, forcing local residents to adapt. Heat waves lead to record temperatures, such as 100.4°F in Siberia in the Arctic Circle in 2020. Wildfires in Australia, the Amazon, California, Oregon, Spain, Portugal, and other areas degrade ecosystems, increase smoke pollution, and displace members of local communities. Together, these extreme weather events provide context for the period of climate instability.

Chapter thesis and organization

Climate change exists as a problem in the era of cascading crises. To address this thesis, the chapter discusses the Hothouse Earth scenario, economics and science of climate change, climate effects, economic and social impacts, the quadruple

---

Case study 4.1 The heat dome and vulnerable members of society

During the summer of 2021, when infections and deaths from the coronavirus pandemic were declining, the northwestern United States was ravaged by an unprecedented heat dome, an "expansive region of high atmospheric pressure characterized by heat, drought and heightened fire danger" (Mann and Hassol, 2021). Weather in the northwest, with warm and cold spells and wet and dry conditions, usually establishes predictable patterns, unless a disturbance occurs. The heat dome, a new climate phenomenon, existed as such a disturbance, developing in the following way: the blazing summer sun first created hot air masses that expanded into the atmosphere. The hot air masses then developed a dome of high pressure, which altered local weather conditions. As high-pressure conditions stabilized, the air dissipated cloud cover and heated the atmosphere. The sun, hotter atmosphere, and decreased cloud cover then heated the ground. Amid drought conditions and a lack of evaporation, hotter temperatures created a feedback loop: the dry landscape intensified the heat dome (Samenow et al., 2021). In Oregon, a record temperature was recorded: 116°F (46°C) in Portland, far exceeding the previous record. But the most vulnerable members of society felt the damage effects most acutely, including the elderly, income-insecure, and homeless, who struggled to survive. In Portland, more than 20 percent of households did not have air conditioning. Poor neighborhoods—with residential towers clustered near interstates and few parks—absorbed and retained heat. A limited number of nighttime cooling centers struggled to provide food, shelter, and medical care. The tragedy of the heat dome with dozens of heat-related deaths demonstrates the impact of global warming if humanity continues to burn fossil fuels at the current rate and temperatures rise over the course of the century.

---

squeeze, and resilience. Readers interested in learning more about the link between the coronavirus pandemic and climate change may read the articles by Crist (2020), Fuller (2020), Hulme et al. (2020), Kantor (2020), and Sengupta (2020), listed in the References section at the end of the chapter.

Hothouse earth

During the summer of 2021, the Earth experienced the hottest month ever on record (July). Canada experienced its hottest day on record, 121°F, in British Columbia. The United States experienced both the hottest temperature in its history and the hottest temperature ever recorded anywhere on Earth, 130°F, in

Death Valley. While these examples provide important data points, the trend for heat waves is clear:

> **Global warming** (the gradual increase in average temperature of the Earth) has caused them to be hotter, larger, longer and more frequent. What were once very rare events are becoming more common. Heat waves now occur three times as often as they did in the 1960s—on average at least six times a year in the United States in the 2010s. Record-breaking hot months are occurring five times more often than would be expected without global warming. And heat waves have become larger, affecting 25 percent more land area in the Northern Hemisphere than they did in 1980.
>
> *(Mann and Hassol, 2021)*

Intergovernmental panel on climate change

According to the Intergovernmental Panel on Climate Change (2021), or IPCC, the world's most important scientific report on climate change, approved by 195 governments and based on 14,000 scientific studies, "It is unequivocal that human influence has warmed the atmosphere, ocean, and land. Widespread and rapid changes in the atmosphere, ocean, cryosphere, and biosphere have occurred." The problem is that high-income countries have delayed fossil fuel abatement. Over the course of the next several decades, the process of global warming will continue.

By burning coal, oil, and natural gas, humans have heated the planet by more than 1°C (2°F) since the nineteenth century. Each of the four decades before the publication of the 2021 IPCC report was warmer than the decade that preceded it. But climatic changes have little historical precedent: "the scale of recent changes across the climate system as a whole—and the present state of many aspects of the climate system—are unprecedented over many centuries to many thousands of years" (IPCC, 2021). In fact, the second decade of this century is "quite likely the hottest the planet has been in 125,000 years" (Plumer and Fountain, 2021). Even if countries reduce greenhouse gas emissions, global warming will continue. With a 1.5°C increase in temperature, relative to pre-industrial levels, dangers grow. Nearly 1 billion people could experience more frequent heat waves, hundreds of millions could struggle with severe droughts, animal species could experience growing levels of extinction, and more extreme weather events could ravage communities (Plumer and Fountain, 2021). But, by the end of the century, the planet will likely experience an average increase in global temperature of at least 2°C beyond pre-industrial levels, even 3°C or 4°C. Each additional degree of warming creates greater perils, including accelerating sea-level rise and vicious floods. Across the planet, heat damage will alter both the natural environment and human civilization.

As global temperatures continue to rise, the IPCC (2021) report explains, so will the hazards. Extreme heat waves that occurred once every half-century will occur once a decade. Tropical cyclones will present annual threats. Ocean levels that are projected to rise by 1–2 feet by the end of the century will flood coastal cities. Further volatile conditions will exist, including unpredictable ocean circulation systems, ecosystem damages, and human displacement: "climate change is already acting in every region, in multiple ways" (Plumer and Fountain, 2021).

The current era

The growing dangers of climate change could soon overwhelm the ability of both human civilization and the environment to adapt. Countries are not doing enough to protect urban areas, agricultural systems, and coastlines from climate hazards: "Rising heat and drought are killing crops and trees, putting millions worldwide at increased risk of hunger and malnutrition, while mosquitoes carrying diseases like malaria and dengue are spreading into new areas" (Plumer and Zhong, 2022). Adverse impacts are becoming more widespread. If temperatures continue to rise, many nations could experience limits in how much they may adapt to changing circumstances. "If nations don't act quickly to slash fossil fuel emissions and halt global warming, more and more people will suffer unavoidable loss or be forced to flee their homes, creating dislocation on a global scale" (Plumer and Zhong, 2022). Despite the growing body of knowledge on climate change, many nations are developing in ways that increase their levels of vulnerability.

Will Steffen of Stockholm University and his coauthors (2018) argue that self-reinforcing feedback in the climate system could push the Earth beyond a planetary **threshold**—a point at which a physical change may occur—causing a Hothouse Earth scenario with disruptions to ecosystems, economies, and migration patterns. According to the authors, a climate emergency characterizes the contemporary global environment. As a result, climate action must address growing problems. While the coronavirus pandemic represents a short-term shock, the threats from climate change, including rising temperatures, increasing wildfires, and extreme drought conditions, will remain for decades.

During the coronavirus pandemic, the vaccination gap between developed and developing countries highlighted the failure of the former to provide aid. More than a year after the onset of the pandemic, the Delta variant, which became a dominant strain of the virus before the Omicron variant appeared, pummeled areas of the world with low vaccination status, including Brazil and India. But the failure to establish an equitable solution with vaccinations served as a reminder of the failure to address climate change. "The vaccine gap presents an object lesson for climate action because it signals the failure of richer nations to see it in their self-interest to urgently help poorer ones fight a global crisis" (Sengupta, 2021).

Economics and science of climate change

The extraordinary heat waves during the coronavirus pandemic, including those that scorched the Pacific Northwest in the United States, were exacerbated by climate change. Temperatures have become so extreme that climate scientists now struggle to determine the rarity of severe events, such as heat domes. A collaborative group of scientists called World Weather Attribution, which works to determine the frequency and magnitude of climate outcomes, warns that, if the world warms by 1.5°C, which will almost certainly occur by the end of the century, heat waves will occur with greater frequency: "The chances of such a severe heat wave occurring somewhere in the world would increase to as much as 20 percent in a given year" (Fountain, 2021a). One reason for concern is that, in many regions, nights are warming faster than days. Hot nights contribute to rising mortality rates, because at-risk individuals, including the elderly, pregnant women, and younger children, do not have a chance to bring their core body temperatures down. A second reason is that temperature records are now broken by a wider margin than ever before. A final reason is that the climate may pass a threshold in which a small rise in average global temperature could increase the likelihood of extreme heat events by a large amount. In other words, the world is facing an immense threat. How immense? A study in *Nature Climate Change* by Drew Shindell of Duke University and his co-authors (2018) argues that "societal risks increase as Earth warms." According to the authors, if the average global temperature increases by 2°C, rather than 1.5°C, 150 million more people may die from air pollution during this century.

Climate history

For most of the Earth's 4.5-billion-year history, planetary conditions were not hospitable. Only in the last 11,500 years have stable climate factors created conditions for the systems—agriculture, government, and markets—of modern civilization. Before that time, temperature changes cycled between ice ages (expanding ice sheets, food shortages, lower sea levels, and water scarcity) and periods of warmth (abundant water, biomass resources, ecosystem diversity, and higher sea levels). In recent millennia, temperature patterns demonstrated little variation. But since 1950, a rapid increase in average global temperature has altered the trajectory of human civilization.

For most of the past 100,000 years, small pockets of humans lived as hunters and gatherers. During periods of climate variability with more difficulty in finding food and shelter, humans were confined to productive savannahs in Africa. In one critical period of cooling, 75,000 years ago, the entire human population may have consisted of 15,000 fertile adults, confined to plateaus in Ethiopia and living close to extinction. By moving into the Arabian Peninsula and then Asia, Europe, and Australia, migrants created semi-nomadic lifestyles.

At the time of 11,500 years ago, the climate stabilized, according to cores drilled into ice sheets in Greenland. During a 40-year period, average global temperatures increased 5°C (9°F), enough to end the last ice age and bring the world into the Holocene period. The Holocene, from 11,500 years ago to 1950, is characterized by expanding ecosystems, a relatively stable climate system, and population growth. During this period, temperatures fluctuated 1°C (2°F) in either direction, creating climate equilibrium, small variation, and stable weather patterns. The outcome was profound: the establishment of modern human civilization (Rockstrom and Klum, 2015).

Anthropocene

Beginning post-World War II and becoming clear by 2020, dramatic increases in global commerce, pollution flows, and urbanization—the movement of the human population from rural to urban areas—have made humans the dominant force for planetary change. The current climate epoch, the Anthropocene, is characterized by population growth, globalization, and environmental degradation. While the first industrial revolution of the mid-eighteenth century, second industrial revolution of the early twentieth century, and third industrial revolution of the late twentieth century created the economic processes that gave rise to the modern era, including steam power, the assembly line, and the digitization of manufacturing, each era relied on fossil fuels. The Anthropocene includes changes in the biosphere, the worldwide sum of all ecosystems, or the zone of life on Earth, that integrates all living organisms and their diversity.

Interconnection

A dynamic form of interconnection exists between the biosphere, atmosphere, and climate system. The circulation of air, flow of water, ocean's conveyer belt, ozone layer, precipitation patterns, soil fertility, glaciers and ice sheets, tectonic plates, volcanic activity, and other factors—both natural and human-induced—shape life in the biosphere. Stability with the worldwide sum of all ecosystems depends on the "complex adaptive interplay between living organisms, the climate, and broader Earth system processes" (Folke et al., 2021). When these factors exhibit increasing levels of variability compared with historical patterns, leading to heat domes, air pollution, and more frequent wildfires, conditions in the biosphere change. These realities reveal an interconnected world. But human activity that creates economic growth also causes climate change, growing inequality, ecosystem transformation, and "calls for transformative change towards sustainable futures" (Folke et al., 2021).

Attribution

Scientists establish links between global warming and severe weather events. But many of the deadly temperature extremes, such as the heat dome in the Pacific Northwest in 2021, "would have been extremely unlikely to occur without human influence on the climate system" (IPCC, 2021). The **rapid attribution analysis** of climate scientists determines the frequency and severity of climate events. But it also establishes links between climate change and specific outcomes, including heat waves, hurricanes, flooding, and droughts. Computer simulations compare what happens in a world of rising global temperatures with a hypothetical world in which human activity has not injected greenhouse gas emissions into the atmosphere for more than two centuries. Using this technique, climate scientists find that the current pace of global warming exacerbates climate outcomes, including heat waves. For example, during the summer of 2021, rapid attribution analysis determined that the heat dome in the Pacific Northwest was "far more likely to occur in the current warmed world than in a world without warming" (Fountain, 2021a).

Climate chaos

As climate scientists explain, a higher atmospheric concentration of greenhouse gases leads to rising temperatures. According to the National Oceanic and Atmospheric Administration, across the planet's land and oceans, the average surface temperature in 2020 was 0.98°C warmer than the twentieth-century average and 1.19°C warmer than the 1900 average (Lindsey and Dahlman, 2021). David Wallace-Wells (2019), author of *The Uninhabitable Earth*, an epoch-defining book on climate change, describes this period of rising temperatures as a period of climate chaos, the extreme alteration of weather patterns:

> The assaults will not be discrete—this is another climate delusion. Instead, they will produce a new kind of cascading violence, waterfalls and avalanches of devastation, the planet pummeled again and again, with increasing intensity and in ways that build on each other and undermine our ability to respond, uprooting much of the landscape we have taken for granted, for centuries, as the stable foundation on which we walk.

Because CO_2 emissions in the atmosphere persist for up to 1,000 years, the actions we take today will continue to impact the climate for generations to come.

Boundaries

Johan Rockstrom and Mattias Klum (2015) use an analogy, guardrails, to describe the idea of **planetary boundaries**, safe operating spaces for humanity. Along highways, guardrails prevent drivers from getting too close to the edge. While guardrails do not hinder the flow of traffic, they reduce accidents. Similarly, planetary boundaries exist to prevent catastrophes, not hinder human development. They are natural processes, dynamic and interrelated, that maintain the planet's operating spaces (Table 4.1).

Each planetary process exists either below the planetary boundary (safe), in a zone of uncertainty (increasing risk), or beyond the zone of uncertainty (high

TABLE 4.1 Planetary boundaries

Planetary boundary	Characteristic
Climate change	High risk
Air pollution	Increasing risk
Chemical flows	High risk
Biodiversity	High risk
Freshwater consumption	Increasing risk
Land-use changes	Increasing risk
Nitrogen and phosphorus pollution	Not yet quantified
Ocean acidification	Increasing risk
Stratospheric ozone	Safe

Source: Rockstrom and Klum (2015).

risk). As one of the nine planetary boundaries, climate change is characterized as high risk, along with chemical flows and biodiversity. Without acting to reduce the risk from climatic changes, the stability of natural systems will decline.

The strength of the planetary boundary model is that it focuses on biophysical processes, including tipping points, which means exceeding the boundaries of system stability. The planetary boundary model defines a safe operating space for humans, demonstrating that, with climate change, biochemical flows, and biodiversity, the world has already gone over the guardrails.

Cost of carbon

Economists define the **social cost of carbon** (SCC) as the economic damage from an additional unit of carbon emissions. This calculation, crucial for climate change policy, provides policymakers with a monetary value for a carbon tax, a per-unit charge on carbon emissions. The idea is to establish a price for carbon emissions, so polluters must decide whether to continue to emit carbon and pay the tax or reduce both carbon emissions and tax payment. Using economics jargon, the polluter will reduce emissions until the marginal cost of emission abatement equals the marginal damage from carbon emissions. At this point, the market internalizes the cost of the carbon emission externality. Policymakers use the SCC in a calculation of the social costs and benefits for policies that involve climate-altering decisions, such as efficiency standards in buildings, fuel efficiency in vehicles, and low-carbon energy sources for power generation. As a monetary value, William Nordhaus (2013) of Yale University, a Nobel-prize winner in the field of economics, uses an estimate of $25 for the SCC. He demonstrates that a carbon tax of this value would increase the price of coal more than oil, natural gas, and electricity, shifting power generation away from its dirtiest source (coal). As environmental economists explain, if the world is to reduce carbon emissions flowing into the atmosphere, establishing a price for carbon emissions would accelerate the process.

Factors causing climate change

Many factors cause the climate to change. Biologists warn that urbanization decimates habitats and ecosystems, increasing extinction rates and weakening the ability of the natural environment to establish carbon sinks in forests and oceans to absorb CO_2 from the atmosphere. They warn that, over the long term, economic development should not create ecosystem damage. Industrialization—the growth of industrial activity—perpetuates fossil fuel consumption in global trade, transportation networks, and power generation. Modernization—the transformation of rural and agrarian societies to urban and industrial centers—locks in climate outcomes. Only major behavioral and policy changes will alter the course of humanity. In this context, what are the effects of climate change? Figure 4.1 offers a perspective.

Climate effects

Droughts

On the planet's surface, 71 percent is covered with water. But barely 2 percent of the total is fresh. Of the freshwater supply, only 1 percent is available for human consumption, while the rest is locked up in glaciers. The outcome is an uneven distribution. Some regions are inundated by monsoons and frequent rainfall. Others are dry. The problem is that a warming planet leads to a greater frequency of drought conditions. Two problems exist. First, because of warm temperatures, desert cities such as Las Vegas, Nevada and Phoenix, Arizona attract new residents. But urban development in these areas increases water scarcity. Second, on a global scale, more than 70 percent of freshwater is used for agriculture and

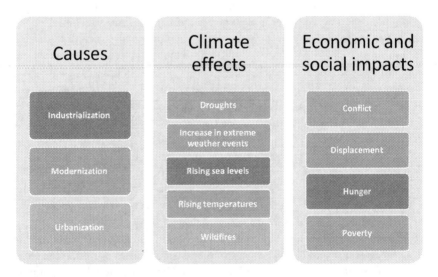

FIGURE 4.1 Causes and effects of climate change.
Source: Author.

irrigation, while more than 10 percent is allocated to industry. In a world of 8 billion people, less freshwater exists for hydration. The world will eventually have 11 billion people, so today's drought conditions will intensify.

The global demand for water outstrips supply. Because of population growth, urbanization, and industrial agriculture, water is made scarce through inefficient infrastructure, policy, and planning. Global warming also contributes. Half of the world's population relies on seasonal melt from snow and ice, accumulations threated by a warming planet. The result is that 2 billion people do not have access to safe drinking water. More than half of the world's population does not have safe sanitation. Over time, global warming will exacerbate these trends.

Increase in extreme weather events

To prophesy the future, humans watch the weather. In a changing climate, the world will experience future weather patterns that include the vengeance of the past. In a 2°C warmer world, the Earth's oceans will warm, drought will exist on a regular basis, and hurricanes, floods, and typhoons will become commonplace. The most important implication of these changes is the need to adapt, a process in which the world becomes better suited to different conditions.

The conflict, displacement, hunger, and poverty that will result from extreme weather events, especially as they impact crop yields and coastal populations, will lead to new forms of organization. Normally, the world is characterized by the slow assembly of nature, the surrounding environment that leads to predictable patterns. But when natural history accelerates, the perspective changes. New flooding, wildfires, and temperatures make previous patterns obsolete. In parts of Siberia, in the Arctic Circle, residents once endured the coldest winters outside of Antarctica. But now they experience wildfires. Warming temperatures in the Russian Arctic feed the blazes that thaw the frozen ground. According to Anton Troianovski (2021), the Moscow bureau chief for *The New York Times*:

> Scientists say that the huge fires have been made possible by the extraordinary summer heat in recent years in northern Siberia, which has been warming faster than just about any other part of the world. And the impact may be felt far from Siberia. The fires may potentially accelerate climate change by releasing enormous quantities of greenhouse gases and destroying Russia's vast boreal forests, which absorb carbon out of the atmosphere.

Some of the world's wildfires exist as predictable characteristics of the natural environment, but many exist now as examples of extreme weather events.

Rising sea levels

As temperature increases and ice and snow on land melts in areas such as Greenland and Antarctica, the result is rising sea levels. Barring a major reduction in greenhouse gas emissions, the estimate is that sea levels could rise between 1 and 2 meters by the end of the century. What will be the impact? Higher sea levels will submerge the Marshall Islands, the Maldives, and many other islands, while threatening coastal areas. The planet is already experiencing these outcomes. In Jakarta, one of the world's fastest-growing cities with a population of more than 10 million people, flooding could leave the entire city under water by the middle of the century. In Lagos, Nigeria's largest city of 15 million people, boasting beach resorts, nightlife, and economic inequality, the rainy season and rising sea level create annual flooding. Miami, with half a million people, could be completely submerged by the end of the century.

With respect to the vulnerability from sea-level rise, cities differ. The factors that determine relative vulnerability include socioeconomic systems, physical infrastructure, urban resources, and local topography (Gargiulo et al., 2020). Around the world, nearly 40 percent of the global population lives within 100 kilometers of the coast. More than half a billion people face a direct threat. For coastal cities, such as Barcelona, Mumbai, New York City, and Tokyo, rising sea levels complicate the process of urbanization.

In response to rising sea levels, coastal cities must adapt. However, even though greater awareness exists on the dangers of rising sea levels, coastal cities have not allocated a sufficient number of resources for adaptation. Without an adequate response, what will submerge is not just the homes of the millions of residents who will flee, but entire neighborhoods, many of which constitute thriving urban areas.

Rising temperatures

Like all mammals, humans are heat-sensitive beings. After experiencing hot temperatures, humans must have the opportunity to cool down. Cooler temperatures, especially at night, draw heat from the skin, so the body keeps pumping. But regions with periods of extreme heat cannot facilitate this function, putting human survival at risk. The reason is that nights are warming faster than days. "And while you might be able to escape the intensifying tropical storms, flooding or droughts by moving elsewhere, refuge from extreme heat is no longer easy to find" (Hassol et al., 2021). With more than 2°C of warming, a reasonable forecast for this century, parts of the equatorial band and hotter latitudes will become unlivable. Because global warming affects all other natural processes, including an increase in extreme weather events and rising sea levels, warmer temperatures lead to an existential threat to current patterns of human existence.

This reality is already delivering wildfires burning ten times more land in Australia, dozens of flooded cities across the world, and regions in India, the Middle East, and the American Southwest that are becoming too hot to experience the outdoors in the summer. "Since 1980, the planet has experienced a fiftyfold increase in the number of dangerous heat waves; a bigger increase is to come" (Wallace-Wells, 2019). In 2003, a European heat wave killed 70,000 people. In 2010, a Russian heat wave killed 55,000 people, 700 a day in Moscow. In 2021, during heat dome conditions, dozens of people died in the Pacific Northwest in the United States.

These examples demonstrate that extreme heat is not just a climate problem but also a public health catastrophe. During hot summer months, Saudi Arabia burns nearly 1 million barrels of oil a day to ensure a stable system of air conditioning. Cities magnify the problems. Asphalt and concrete, the materials of urban density, absorb so much ambient heat during the day that the release of it at night can raise nighttime temperatures to uncomfortable levels, eliminating the possibility of proper cooling. Over time, with rising temperatures, regions such as northern Minnesota and Wisconsin will experience an increase in demand for housing.

Wildfires

As the increasing number of wildfires in California, Oregon, Australia, the Amazon, and other parts of the world demonstrate, climate change exacerbates the environmental conditions that give rise to conflagrations. In one example, with the Bootleg Fire in Southern Oregon in 2021, the largest fire in the country during the year, an intense heat wave and months of drought fueled the fire that burned more than 500 square miles of grassland and forest. Flaming for weeks, the Bootleg Fire exhibited the characteristics of an extreme weather event, igniting stands of trees, altering wind patterns, and leaping fire barriers. It even gave rise to the terms fire tornado, "swirling vortexes of heat, smoke and high winds," and fire whirls, "small spinning vortexes of air and flames" (Fountain, 2021b). The heat stemmed from the size of the fire and the dryness of the vegetation. It was so extreme that dozens of homes burned. Firefighters had to retreat.

The Bootleg Fire burned for weeks; however, it was not unique. It was similar to many other intense wildfires, including the Camp Fire in California in 2018. When conditions become hotter and drier from climate change, wildfire activity increases. Because forests are drier from increased evaporation, these ecosystems are primed for fires to ignite and spread. The result is longer, costlier, and more extreme wildfires. In some parts of the world, such as Australia and California, wildfires burn all year.

Economic and social impacts

Conflict

Climate effects, especially as they relate to more extreme weather conditions, lead to drought and stress on the land, especially in areas suffering from drier conditions. As a result, even small increases in average global temperature will lead to a greater risk of human conflict. Consider the reasons. In poorer regions of the world where individuals rely on the land, reduced crop yields and stress on freshwater aquifers lead to resource scarcity. Resource scarcity increases competition for the means of survival, enhancing the potential for conflict. In areas such as Somalia, Sudan, and Syria, resource scarcity, climate change, and conflict are intertwined. In this context, risk factors exist: extreme weather events, livelihood insecurity and migration, local resource competition, sea-level rise, transboundary water management, volatile food prices, and unforeseen effects (Busby, 2021). The most severe threat is the loss of life. On a larger scale, climate change constitutes an ongoing security concern. But the extent to which the loss of livelihoods from climate hazards rises to the level of national security depends on resource scarcity and the ability to assimilate the displaced.

Displacement

With respect to devastating consequences, one of the most important is displacement. Climate refugees, individuals displaced from their homes because of severe weather patterns, are moving away from some of the poorest parts of the planet. But climate refugees also exist in developed countries. Even in richer parts of the world, rising sea levels will submerge coastal cities. In desert communities, droughts will eliminate local water supplies, forcing residents to leave. Hotter temperatures are pushing some areas to the brink. In 2020, Phoenix, Arizona endured 53 straight days of 110°F (43°C) heat, shattering the record by 20 days. This extreme heat wave placed undue burden on the city's energy system to provide air conditioning, leading to additional greenhouse gas emissions from coal-fired power plants.

With climate change, this pattern will continue, but it will impact poorer countries the most: "In much of the developing world, vulnerable people will attempt to flee the emerging perils of global warming, seeking cooler temperatures, more fresh water and safety" (Lustgarten, 2020). In developed countries, people may migrate to unstable regions, settling in cloudless deserts or coastlines. The world, in other words, is on the cusp of transformation. One estimate is that, in the United States, "162 million people—nearly 1 in 2—will most likely experience a decline in the quality of their environment, namely more heat and less water" (Lustgarten, 2020). Over the course of this century, many people will move, especially those living in places such as Phoenix and Las Vegas, outside the ideal niche for human existence, experiencing the dual threat of heat and drought. In upcoming decades, the cost of human displacement will rise.

The monetary outlays necessary to defend neighborhoods against rising sea levels, pipe water hundreds or thousands of miles to parched cities, and establish new food supply chains may become cost-prohibitive. But climate threats are too expensive to ignore. In both developed and developing countries, shifting populations increase poverty, test the ability of public sectors to supply basic services, amplify inequalities, and burden the most vulnerable members of society, a recurring theme throughout this book. While mobility is a function of income and wealth, those who cannot move may become trapped, as society struggles to offer support.

Hunger

Hundreds of millions of people face an inadequate food supply. But an increase in temperature will make the problem worse. Adapting to rising temperatures will place additional burden on both food production and distribution; however, uneven outcomes will exist. The livelihoods of pastoral people, subsistence farmers, and those who rely on local food production are most sensitive to climate effects. In developed countries, those who live in food deserts will continue to face unbalanced access to healthy calories. Water scarcity decreases the stability of food supply chains, altering health, nutrition, and security. In agricultural systems, even with a small increase in temperature, crops will reach their maximum tolerance, especially where non-irrigated land exists. Drier latitudes with declining crop yields will experience lower levels of agricultural productivity. As it relates to hunger, however, climate change will serve as a driver of inequality, as extreme weather events disproportionately impact the world's poorest people.

An article in *Nature Climate Change*, addressing climate change, hunger, and trade networks, argues that almost 1 billion people currently suffer from hunger (Janssens et al., 2020). The article estimates that, by 2050, with current levels of trade integration, climate change will increase the number of undernourished people by millions. If greater levels of trade integration link food-deficit regions with food-surplus regions, however, consumption possibilities could increase in vulnerable areas, creating a framework for adaptation. The key in the process is to establish resilient food networks, reduce the costs of trade, and employ sustainable methods of farming and distribution that minimize greenhouse gas emissions.

Poverty

The effects of climate change, including conflict, displacement, and hunger, disproportionately impact the most vulnerable members of society. In this context, migration often exists as an economic phenomenon. Individuals move to seek better opportunities. But people may also flee from their places of residence for reasons of political persecution, social inequities, and environmental degradation. One climate change outcome is therefore forced migration, when people leave areas of drought, resource scarcity, and poverty. An inadequate supply of basic resources complicates the establishment of methods of subsistence living. At the community level, climate change and poverty are intertwined. Poverty serves as a driver of vulnerability to climate effects. Individuals and communities with fewer resources experience lower capacities to implement adaptive responses. Over time, the ability of the most vulnerable members of society to address the problems of climate change and poverty will depend on informed policy, demographic changes, and methods of resilience.

Quadruple squeeze

In a study of climate trends, Rockstrom and Klum (2015) identify four factors—climate variability, ecosystem changes, population growth, and

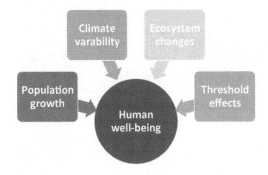

FIGURE 4.2 Quadruple squeeze.
Source: Adapted from Rockstrom and Klum (2015).

threshold effects—that constitute the quadruple squeeze, factors that impact human well-being on a large scale (Figure 4.2).

Population growth

It took 2 million years of human history for the global population to reach 1 billion, in 1804, during the first industrial revolution. It took 200 more years to reach 7 billion, in 2011, during the fourth industrial revolution, with the world increasing to 8 billion a decade later. But for the first time in human history, global population is expected to plateau, stabilizing at 11 billion at the end of the century. But most of the increase will occur in developing countries. Stabilization will result from a declining fertility rate, currently 2.4 children per woman, forecasted to fall below two by 2100. The reason is more education and economic opportunity for women. But even when population stabilizes, a world with 11 billion people will require almost a 50 percent increase in food production, relative to global production in 2020, as living standards in developing countries increase. In this future position, individuals living in stable economic, political, and social systems with a greater access to resources will maintain higher living standards than those who do not.

Climate variability

For decades, climate scientists have analyzed the role of CO_2 in warming the planet. Writing in *The Atlantic*, Peter Brannon (2021) describes the dynamic relationship:

> We live on a wild planet, a wobbly, erupting, ocean-sloshed orb that careens around a giant thermonuclear explosion in the void. . . . Of more immediate interest today, a variation in the composition of the Earth's atmosphere of as little as 0.1 percent has meant the difference between

weltering Artic rainforests and a half mile of ice atop Boston. That negligible wisp of the air is carbon dioxide.

During the planet's history, large levels of CO_2 have leaped in natural processes from the seas and risen from the crusts, warming the planet. During other periods, CO_2 has hidden in the ocean depths and rocks, cooling the planet. But during the entire half-billion-year period of animal life, "CO_2 has been the primary driver of the Earth's climate" (Brannon, 2021). With the exception of the coronavirus pandemic and recessionary intervals, CO_2 emissions have increased during expansionary phases of the economy, stemming from industrial agriculture, manufacturing, power generation, and transportation (Figure 4.3).

The reason for rising temperatures is the link between carbon emissions and the atmospheric concentration of CO_2. This link gives rise to the **greenhouse effect**, the reason that the Earth is hospitable to life. The greenhouse effect refers to the condition in which solar radiation from the sun that passes through the atmosphere is absorbed and re-emitted in all directions by greenhouse gas molecules. The effect is a warmer atmosphere. But greenhouse gases emitted from human activity increase average global temperatures, disrupting the stability of the climate system. Throughout history, when there has been as much CO_2 in the air as today, the planet has been a warmer place, with oceans as much as 70 feet higher. When the atmospheric concentration of CO_2 rises, average global temperatures increase. As recently as 1970, CO_2 in parts per million (ppm) equaled 325. But in 2020, it rose to 414, an unprecedented level in recent history (Figure 4.4). The forecast is for the atmospheric concentration of CO_2 to continue to rise.

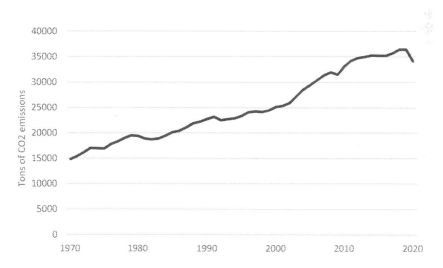

FIGURE 4.3 Global CO_2 emissions.

Source: Author using data from Our World In Data, https://github.com/owid/co2-data

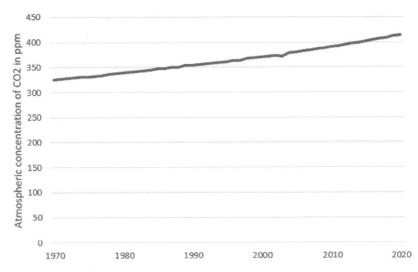

FIGURE 4.4 Atmospheric concentration of CO_2.

Source: Author using data from the National Oceanic and Atmospheric Administration, https://gml.noaa.gov/webdata/ccgg/trends/co2/co2_annmean_mlo.txt

Ecosystem changes

Ecosystems—areas of the environment that sustain the natural world and pro-vide economic benefits—include coral reefs, inland wetlands, lakes and rivers, mangroves, tropical forests, woodlands and shrubs, and grassland. But deforesta-tion in the Amazon to clear land for cattle degrades the world's most important tropical forest. While short-term land lease agreements and a lack of enforcement of existing property rights contribute to the problem, the largest incentive is the world's increasing demand for meat. By clear-cutting forests and reducing the capacity of the trees to absorb CO_2, the action contributes to climate change. In general, as humans degrade the air, forests, soils, and waterways, ecosystem services decline, leading to a decrease in human well-being. Because of the cur-rent scale of destruction, it is important to draw attention to the problem. But as the environment loses its ability to overcome human incursions, the natural world will not be able to accommodate pollution, chemical emissions, and other degrading flows. It will become even more difficult to sustain ecosystem services, such as clean air and water, which are necessary for a healthy planet.

Threshold effects

Upper limits provide bounds for normal temperature patterns, depending on historical rates of warming, previous records for heat, and climate variability. But standard climate models suggest that recent trends, especially with extreme tem-peratures, should not be possible. According to a study in *Scientific American*, the record increase in temperatures during the heat dome conditions in the Pacific Northwest in the United States in 2021 would have been "150 times less likely in a world without climate change" (Harvey, 2021). It may be the case that the flow of the jet stream or other climate condition contributed to this extreme weather event. But it may also be that the region crossed a threshold, a point at which the natural environment creates new conditions, flows, and patterns. After passing a threshold, weather events that were considered extreme become normal. An ecological system, such as a coral reef, has a built-in capacity to withstand an external threat, such as rising ocean temperatures. But the more pronounced is the threat, the larger is the possibility that the system will lose its ability to adapt. Thresholds exist because natural system feedbacks either maintain equilibrium conditions or amplify perturbations. Some of the feedbacks that are weakening the climate system, including an increase in carbon emissions and a rise in atmospheric concentration of CO_2, may drive the climate system to a different state with new equilibrium conditions and variations. After a threshold is crossed, new feedbacks become self-perpetuating. Events previously considered extreme characterize new conditions, weakening the ability of natural systems to remain resilient (Steffen et al., 2018).

Resilience

A problem with climate change is that, even if carbon emissions decline, heat waves and other extreme weather events will continue, because so much CO_2 is locked into the atmosphere. This reality constitutes the reason why human civilization must increase its level of resilience—the capacity to recover from extreme levels of change—in a world of rising temperatures. What does the future hold? In the absence of behavioral changes that reduce fossil fuel consumption, higher temperatures will continue to create dangerous climate outcomes. On a global scale, 5 million people die annually from excessive heat, a total that will likely increase.

To establish a position of resilience, countries must develop early-warning systems, heat action plans, programs of clean-energy transformation and decarbonization, and power grid improvements. With these plans, heat-related disruptions may not become life-threatening. But countries must remember that climate effects will disproportionately affect individuals with chronic illnesses, the elderly, laborers who work outdoors, people living in poverty, and those with mobility problems and conditions of isolation.

Retreat

After experiencing multiple floods, residents of the town of Valmeyer, Illinois decided in 1993 to choose a radical path. Instead of allocating scarce resources to fight a never-ending battle to hold back floodwaters from the Mississippi River, members of the community decided to move the entire town to higher ground.

Using funds from the state of Illinois and the Federal Emergency Management Agency, they moved the town a few miles away.

As flooding continues in wetter environments and wildfires become more intense in drier regions, more communities will act like Valmeyer. Opting for a path of **managed retreat**, they will view this choice as a cost-effective option. In an article in *Science*, Katherine Mach and A.R. Siders (2021) argue that, even though managed retreat includes the economic and psychological costs of moving, it offers a future pathway. With displacement from climate change, unplanned retreat is already happening, as individuals move away from environmental hazards. But with managed retreat, relocation occurs safely, preserving economies and social justice in a wider range of strategies. In some cases, managed retreat entails targeted efforts, such as rerouting roads, moving homes, or creating more space for water pumps. In other cases, managed retreat entails the relocation of an entire community. Either way, it provides an opportunity for a better future.

The need for collective action

Climate change and pandemics threaten lives and livelihoods, requiring collective action, functional public sectors, and scientific expertise. With both crises, individual action does not solve society's problems. With the pandemic, accurate information on the number of confirmed cases informs intervention measures. With climate change, accurate information on the impact of a warming planet informs policy responses. "Humans are part of nature, not separate from it, and human activity that hurts the environment also hurts us" (Crist, 2020). Climate change and disease outbreaks will create future challenges, including the best way to adapt to instability.

Adaptation, a dynamic social process, is necessary in an unstable world. In this context, social capital, the networks of relationships among people, determines the extent to which society addresses a challenge, such as climate change or a pandemic. Social capital frames "both the public and private sector institutions of resource management that build resilience in the face of risks" (Adger, 2003). With climate change and pandemics, the key element for a society's response is collective action. But when dissenting voices or political predilections encourage a distrust of scientific evidence, such as vaccinations helping to solve a pandemic or human activity serving as the main cause of climate change, society may struggle to solve impending problems.

Moving forward

Adapting to climate threats in a world with billions of people requires innovations in our methods of commerce, distribution, energy, farming, globalization, and urban design. In this world, climate change requires a coordinated and multi-generational commitment, especially from developed countries, that will alter every part of society. While sea walls reduce the impact of coastal hazards and social solutions address conflict and displacement, nature-based solutions enhance natural systems, providing additional levels of support. Nature-based solutions, which encompass actions that "protect, restore, or sustainably manage ecosystems," include ecosystem maintenance, sustainable urban infrastructure, and landscape and forest restoration (Chausson et al., 2020). These actions entail biodiversity preservation, cleaner air, emission reduction, flood control, and urban cooling. An important result is the impact in multiple areas.

Overall, the coronavirus pandemic and climate change provide points of connection. They have global impacts. They require cooperation. Most importantly, they are "problems of exponential growth against a limited capacity to cope," according to Elizabeth Sawin, codirector of Climate Interactive (Gardiner, 2020). By infecting a large number of people in a short period of time, a pandemic overwhelms healthcare systems. But the growth in carbon emissions overwhelms society's ability to manage hotter temperatures, conflict, displacement, hunger, and poverty. While a pandemic impacts the global population faster than climate change, both require collective action. While the world waits for an appropriate response, people suffer.

Summary

Climate change, an increase in extreme weather events such as global warming, more frequent and intense wildfires, drought, and sea-level rise, stems from globalization, modernization, and urbanization. These latter processes, which increase economic activity, rely on fossil fuels, including coal, oil, and natural gas. The outcomes of the fossil fuel era include rising carbon emissions and an increase in the atmospheric concentration of CO_2. Hotter temperatures present economic, environmental, and social challenges that will occupy the world for decades. Links exist between climate change and the coronavirus pandemic. Climate change facilitates pandemics. Pandemics alter the process of climate change. Over time, as climate effects lead to greater levels of conflict, displacement, hunger, and poverty, the world will have to identify strategies of adaptation, collective action, decarbonization, and managed retreat.

Chapter takeaways

LO1	The coronavirus pandemic and climate change interact in harmful ways.
LO2	The Hothouse Earth scenario entails higher average global temperatures, more extreme heat waves, and rising damage effects.
LO3	Urbanization, industrialization, and modernization contribute to a changing climate.
LO4	Climate effects include drought, an increase in extreme weather events, rising sea levels, rising temperatures, and wildfires.

LO5   Economic and social impacts of climate change include conflict, displacement, hunger, and poverty.

LO6   The quadruple squeeze entails climate variability, ecosystem changes, population growth, and threshold effects.

LO7   Strategies of resilience include managed retreat, collective action, and sustainable future pathways.

Key terms

Carbon dioxide   Managed retreat

Climate change   Planetary boundaries

Decarbonization   Rapid attribution analysis

Global warming   Social cost of carbon

Greenhouse effect   Threshold

Greenhouse gas emissions

Questions

1   How does climate change relate to the coronavirus pandemic?

2   In the context of climate history, how would you characterize the current climate period?

3   What are the most important drivers of climate change?

4   Which climate effects will create the most damage to humans and the environment?

5   Considering the economic and social impacts of a changing climate, including conflict, displacement, hunger, and poverty, which are most pronounced on a global scale?

6   Explain the quadruple squeeze. In the model, what is the relative degree of importance of the climate emergency?

7   What methods of climate adaptation are the most effective?

8   Does a framework of resilience apply to climate change and pandemics? Explain.

References

Adger, W. Neil. 2003. "Social capital, collective action, and adaptation to climate change." *Economic Geography*, 74(4): 387–404.

Beyer, Robert, Manica, Andrea, and Mora, Camilo. 2021. "Shifts in global bat diversity suggest a possible role of climate change in the emergence of SARS-CoV-1 and SARS-CoV-2." *Science of the Total Environment*, 767: 145413.

Brannon, Peter. 2021. "The Dark Secrets of the Earth's Deep Past." *The Atlantic*, March, 60–75.

Busby, Joshua. 2021. "Beyond internal conflict: the emergent practice of climate security." *Journal of Peace Research*, 58(1): 186–194.

Chausson, Alexandre, Turner, Beth, Seddon, Dan, ...Seddon, Nathalie. 2020. "Mapping the effectiveness of nature-based solutions for climate change adaptation." *Global Change Biology*, 26(6): 6134–6155.

Crist, Meehan. 2020. "What the Pandemic Means for Climate Change." *The New York Times*, March 29.

Editorial. 2022. "A World on Fire." *The New York Times*, January 2.

Folke, Carl, Polasky, Stephen, Rockstrom, Johan...Walker, Brian. 2021. "Our future in the Anthropocene biosphere." *Ambio*, 50: 834–869.

Fountain, Henry. 2021a. "Scientists Say Extreme Heat Is a Grim Sign." *The New York Times*, July 8.

Fountain, Henry. 2021b. "How Bad Is the Bootleg Fire? It's Generating It's Own Weather." *The New York Times*, July 19.

Fuller, Thomas. 2020. "A Wary California Released Inmate Firefighters. Now, Fires Rage." *The New York Times*, August 23.

Gardiner, Beth. 2020. "Coronavirus Holds Key Lessons on How to Fight Climate Change." *Yale Environment 360*, March 23.

Gargiulo, Carmela, Battara, Rosaria, and Tremiterra, Maria. 2020. "Coastal areas and climate change: a decision support tool for adaptation measures." *Land Use Policy*, 91: 104413.

Harvey, Chelsea. 2021. "Western Heat Wave 'Virtually Impossible' without Climate Change." *Scientific American*, July 8.

Hassol, Susan, Ebi, Kristie, and Serkez, Yaryna. 2021. "America in 2090: The Impact of Extreme Heat, in Maps." *The New York Times*, July 21.

Hulme, Mike, Lidskog, Rolf, White, James, and Standring, Adam. 2020. "Social scientific knowledge in times of crisis: What climate change can learn from coronavirus (and vice versa)." *Wiley Interdisciplinary Reviews: Climate Change*, e656: 1–5.

IPCC. 2021. *Climate Change 2021:* "Contribution of working group I to the sixth assessment report of the intergovernmental panel on climate change." In Masson-Delmotte, V., Zhai, P., Pirani, A... (Eds.), *The Physical Science Basis*. Cambridge: Cambridge University Press.

Janssens, Charlotte, Havlik, Petr, Krisztin, Tamas...Maertens, Miet. 2020. "Global hunger and climate change adaptation through international trade." *Nature Climate Change*, 10: 829–835.

Kantor, Marianna. 2020. "On Coronavirus and Climate Change." *Forbes*, July 17.

Kiser, Daniel, Elhanan, Gai, Metcalf, William, Schnieder, Brendan, and Grzymski, Joseph. 2021. "SARS-CoV-2 test positivity rate in Reno, Nevada: association with PM2.5 during the 2020 wildfire smoke events in the western United States." *Journal of Exposure Science & Environmental Epidemiology*, 31: 797–803.

Lindsey, Rebecca and Dahlman, LuAnn. 2021. "Climate Change: Global Temperature." *National Oceanic and Atmospheric Administration*, March 15.

Lustgarten, Abraham. 2020. "Climate Change Will Force a New American Migration." *ProPublica*, September 15.

Mach, Katherine and Siders, A.R. 2021. "Reframing strategic, managed retreat for transformative climate adaptation." *Science*, 372(6548): 1294–1299.

Mann, Michael and Hassol, Susan. 2021. "That Heat Dome? Yeah, It's Climate Change." *The New York Times*, June 29.

Nordhaus, William. 2013. *The Climate Casino: Risk, Uncertainty, and Economics for a Warming World*. New Haven, CT: Yale University Press.

Plumer, Brad and Fountain, Henry. 2021. "A Hotter Future Is Certain, Climate Panel Warns. But How Hot Is Up to Us." *The New York Times*, October 7.

Plumer, Brad and Zhong, Raymond. 2022. "Climate Change Is Harming the Planet Faster Than We Can Adapt, U.N. Warns." *The New York Times*, February 28.

Rockstrom, Johan and Klum, Mattias. 2015. *Big World Small Planet*. New Haven, CT: Yale University Press.

Samenow, Jason, Galocha, Artur, and Leonard, Diana. 2021. "The Science of Heat Domes and How Drought and Climate Change Make Them Worse." *The Washington Post*, July 10.

Sengupta, Somini. 2021. "Global Vaccine Crisis Send Ominous Signal for Fighting Climate Change." *The New York Times*, May 4.

Sengupta, Somini. 2020. "Heat, Smoke and Covid Are Battering the Workers Who Feed America." *The New York Times*, August 25.

Shindell, Drew, 2018. "Quantified, Localized Health Benefits of Accelerated Carbon Dioxide Emissions Reduction." *Nature Climate Change*, 8: 291–295.

Steffen, Will, Rockstrom, Johan, Richardson, Katharine, …, Schellnhuber, Hans. 2018. "Trajectories of the earth system in the Anthropocene." *PNAS*, 115(33): 8252–8259.

Troianovski, Anton. 2021. "As Frozen Land Burns, Siberia Fears: 'If We Don't Have the Forest, We Don't Have Life.'" *The New York Times*, July 17.

Wallace-Wells, David. 2019. *The Uninhabitable Earth: A Story of the Future*. New York: Penguin Press.

PART II
Social instability

5

RACIAL INJUSTICE

Chapter learning objectives

After reading this chapter, you will be able to:

LO1   Identify the convergence between racial injustice and the coronavirus pandemic.

LO2   Recognize systems of privilege and inequality.

LO3   Analyze antiracism as a transformative concept.

LO4   Establish a framework of antiracism intervention.

LO5   Discuss the connection between racism and police brutality.

LO6   Address the problem of residential segregation.

LO7   Consider the implications of environmental racism.

LO8   Explain the inequitable outcomes of healthcare systems.

Chapter outline

Protect and unite
Systems of privilege and inequality
Antiracism
Intervention and reform
Police brutality
Residential segregation
Environmental racism
Health inequities
Summary

DOI: 10.4324/9781003310075-7

Protect and unite

If the coronavirus pandemic demonstrated how unresponsive public sectors may fail to protect the public, the brutal murder of George Floyd in Minneapolis, on May 25, 2020, while the pandemic raged, showed how public sectors may do harm. The 46-year-old African American, handcuffed, was on the ground for more than 9 minutes as a police officer knelt on Floyd's neck until he died. For many who watched the video, it will be forever ingrained in their minds. During a time of continued incidents of police brutality, the George Floyd tragedy led to massive and recurrent protests throughout the world.

Early in the pandemic, warnings about the novel coronavirus were inconsistent; however, in the United States, warnings about policing were historical and repetitive. More than 3 decades after the Rodney King beating in Los Angeles led to calls for reform, in 1991, police in the United States during the third decade of this century are three times more likely to kill African Americans than white people. For young Black men, death-by-cop serves as the sixth leading cause of death. To this day, in the United States, "police violence persists unabated" (Smith, 2021). Many more African Americans than white people serve time in prisons, even though the former are one-sixth of the population of the United States (Micklethwait and Wooldridge, 2020).

In the United States, two constitutional problems bedevil police reform. First, because of the second amendment—the right to bear arms—the country is heavily armed. Per 100 individuals in the United States, 120 civilian firearms exist. As a result, in the country, almost 15,000 people die annually in gun homicides. In urban areas, many police officers are terrified of being shot. Many of the victims of police shootings are armed. Toughening up background checks and removing guns from the streets serve as important policy goals. But the federal government should also eliminate the Pentagon program that distributes surplus weapons to police: "When police swagger around even small towns with armed Humvees and machine guns that have seen service in Iraq, they look like an occupying army" (Micklethwait and Wooldridge, 2020). Second, in the United States, more than 18,000 law enforcement agencies exist. While small towns require a specific area of oversight, in larger cities, such as Los Angeles or Minneapolis, multiple forces overlap, complicating enforcement.

The definitions

In his book, *How to be an Antiracist*, Ibram Kendi (2019), Professor of History and International Relations at American University, contrasts **antiracist** with **racist** sentiments. An antiracist is "One who is supporting an antiracist policy through their actions or expressing an antiracist idea." An antiracist idea is one that means racial groups are "equals in their apparent differences—that there is nothing right or wrong with any racial group." In contrast, a racist, in Kendi's framework, is "One who is supporting a racist policy through their actions or inaction or expressing a racist idea." A racist idea is one that means racial groups are not equal in their apparent differences.

For antiracists, the act of supporting antiracist policy entails the actions that combat problems of racism. According to Kendi (2019), antiracists believe social problems are rooted in "power and policies," racists believe social problems are "rooted in groups of people," and racist actions lead to racial inequity, "when two or more racial groups are not standing on approximately equal footing."

In the United States, home ownership serves as an example. According to the Census Bureau, in 2019, 65 percent of families lived in owner-occupied homes. Informational categories, however, tell a different story. With respect to white households, 73 percent lived in owner-occupied homes, compared with 51 percent for Indigenous or Alaskan native households, 48 percent for Latinx households, and 42 percent for African American households. Why does this difference exist? According to one study, differences in credit scores and income help to explain the home ownership gap (Choi et al., 2019).

An antiracist policy creates or perpetuates racial equity, while a racist policy creates or perpetuates racial inequity. In this context, policies include the "unwritten laws, rules, procedures, processes, regulations, and guidelines that govern people" and every policy in every institution is "producing or sustaining either racial inequity or equity between racial groups" (Kendi, 2019). With the examples in this chapter, including police brutality, residential segregation, environmental racism, and health inequity, the definitions of antiracism and racism inform the discussion, highlighting policy outcomes.

Racial injustice: an element of the cascading crises

During the coronavirus pandemic, fear about the spread of the pathogen created widespread limits on human contact. At the same time, an economic shutdown forced businesses to close, employees to choose remote options, and frontline workers to keep the economy running. But the killing of Black women and men during the pandemic—Ahmaud Arbery, Breonna Taylor, and George Floyd, among others—created widespread unrest that led to repeated demonstrations in the United States and abroad. Rallies for racial justice called for reforms in a U.S. system of policing that, for centuries, disproportionately harmed people of color. But other examples characterized the period, including acts of violence against Asian Americans, environmental racism with Indigenous people, and residential segregation that compounded the spread of the virus.

This chapter acknowledges the intersection of multiple effects. For example, individuals in marginalized communities may experience relatively higher levels of susceptibility to a new pathogen, inferior health outcomes, environmental racism, residential segregation, and unemployment. Although these examples demonstrate intersectional realities, it is important to acknowledge that, with marginalized communities, inequitable policy outcomes may persist. As the chapter argues, the implication is that racism is complicated, historical, and nuanced, intersecting with divisions in the socioeconomic hierarchy.

An analysis of the era of cascading crises reveals that problems of racial injustice persisted. Because of Covid-19, millions of people worldwide lost their lives, while vulnerable members of the population were impacted in a disproportionate manner. The number of unemployed grew to tens of millions of people, but higher rates of unemployment existed in communities of color. For marginalized groups, a central narrative entailed the role of underlying divisions that shaped human outcomes, whether in the realms of the economy, health, or society. Sandro Galea and Salma Abdalla (2020) of Boston University, writing during the period, characterized the sentiment:

> The resurgence of anger at long-standing racism and racial inequities was added to the anxiety and tension of the pandemic, creating a combustible scene of national civil unrest. Deep political divisions have shaped the moment from the start. Partisan divides have informed opinions around the extent of a national shutdown needed to mitigate pandemic spread, a pandemic that has disproportionately led to the deaths of black people, and about how to address the legitimate concerns of thousands of individuals protesting the murder of black men and women.

Chapter thesis and organization

The chapter's thesis is that, in a time of anxiety, instability, and uncertainty, racist policies exacerbate inequitable outcomes. To address the thesis, this chapter first describes systems of privilege and inequality. To consider why antiracism serves as a transformative concept that reenergizes the contemporary debate, the chapter then discusses Ibram Kendi's (2019) book, methods of intervention, and policy reform. Using this organizing framework, the chapter considers racial injustice during the coronavirus pandemic: police brutality, residential segregation, environmental racism, and inequitable health outcomes.

Systems of privilege and inequality

Individuals of any ethnicity, gender, or race are as different as they are alike. Although they share elements of their existence, including socializing, taking care of children, or working outside the home, their lives are characterized by difference. Differences stem from cultural norms, institutional influences, material practices, and political conditions. For example, individuals from the same group may experience formal or informal sectors, religious or nonreligious beliefs, traditional or nontraditional values, urban or rural geographies, and western or eastern cultures. Expectations dictate patterns of behavior and impose sanctions when accepted behaviors are broken. Choices create multifaceted outcomes, including discrimination, expectations, power, privilege, respect, and the value of difference. This reality demonstrates what it means to be an individual in a society with multiple identities and complex interactions (Shaw and Lee, 2012).

Hierarchy and difference

Through behavioral practices, cultures and values, economic opportunities, and material conditions, societies illustrate differences. But, for individuals, the intersection of ability, age, background, ethnicity, race, sexual identity, and status determines identity. A national context then situates identity within a social order. With colonialism, imperialism, and oppression, the social order subordinates some individuals to others. Cultural expectations, market organization, and social roles initiate, influence, and maintain the social order.

This framework captures the socially constructed nature of difference. A category such as "young adult," for example, describes individuals among a certain age group but with different backgrounds, beliefs, ethnicities, genders, practices, and races. Society may characterize an "urban" young adult with one set of characteristics and a "rural" young adult with another set of characteristics. Society may not view a "female" young adult the same way as a "male" young adult. Discrimination results from the notion that individuals should behave, decide, and look in culturally specific ways. Taken together, these realities establish a context of hierarchy and difference, according to Shaw and Lee (2012):

> Society recognizes the ways people are different and assigns group membership based on these differences; at the same time, society also ranks the differences and institutionalizes them into the fabric of society. Institutionalized means officially placed into a structured system or set of practices. In other words, institutionalized means to make something part of a structured and well-established system.

Institutionalization

The implication of hierarchy and difference is that attitudes, practices, and social norms may exclude individuals of certain ethnicity, gender, and/or race from specific roles in society, such as law enforcement, leadership, or management. These areas may contain cultured content, responsibilities, and social expectations that are less accessible to certain individuals, such as men teaching in elementary schools. "The concept of institutionalization . . . implies that meanings associated with difference exist beyond the intentions of individual people" (Shaw and Lee, 2012). Because of the existence of systems of privilege and inequality, individuals may associate the concept of "leaders," "police officers," or "teachers" with attributes of ethnicity, gender, and race; however, these characteristics describe the individuals who occupy the positions. Individuals may envision an elementary school teacher to be a certain combination of ethnicity, gender, and race, but in fact many different kinds of people serve as elementary school teachers. Within the social order, interaction with leaders, police officers, and teachers creates beliefs, expectations, and practices. Over time, these beliefs, expectations, and practices become normalized. As a result, breaking historical patterns exists as a difficult and multifaceted process.

Institutions of society, the organizations of specific purpose that establish patterns of behavior—economy, education, government, law enforcement, public health—are historical, meeting the needs of the dominant group. The dominant group creates the institutions and their organizing purposes. The result is both inequality and privilege, as institutions enforce positions of domination and subordination (Shaw and Lee, 2012). Those who hold power influence those who do not, a reality that helps to explain why racist policies persist.

This chapter addresses racial injustice during the era of cascading crises. But systems that facilitate inequality and privilege include several factors, not just race, that intersect to create bias, discrimination, and stratification. Additional factors include age, ethnicity, gender, language, place of origin, religion, sexual preference, and socioeconomic class. With these factors, individuals of the same ethnicity, gender, race, or other characteristic experience the world through unique lenses, depending on their positions within the socioeconomic order. A 10-year-old growing up in the slums of Mumbai, India has a different view of the world than a 10-year-old growing up in the wealth of Greenwich, Connecticut.

Paradigms of race

In the book *Racial Formation in the United States*, Michael Omi and Howard Winant (2015) argue that "race and racial meanings are neither stable nor consistent," that is, contradictions concerning race abound in contemporary society, as they did in the past. According to the authors, "racial inequalities pervade every institutional setting." As a result, an individual's racial identity may differ from society's perception of racial identity. In this framework, race exists as a social construct, a category of agency, difference, and inequity (as does ethnicity and gender). Omi and Winant (2015) frame race on the basis of ethnicity, class, and nation.

Ethnicity

Ethnicity-focused theories of race argue for assimilation, cultural pluralism, and inclusion. Their ascent in the early- and mid-twentieth century, affording primacy to cultural values, led to their decline in the late twentieth century. "Ethnicity theory was in fact the first mainstream social scientific account of race to understand it as a socially constructed phenomenon" (Omi and Winant, 2015). In the early twentieth century, the idea of ethnicity-focused theories of race was associated with the influx of European immigrants into the United States. This process created the need to assign identity and social status to new groups of people.

As a challenge to biologistic depictions of race, ethnicity theories operated in cultural contexts but were limited by their applications. Only after World War II did the idea move from analysis of the U.S. racial frontier, an imperialist meaning, to racial otherness. "As long as race could be subsumed under the ethnicity label . . . the immigrant analogy could be applied" (Omi and Winant, 2015). In the 1950s and 1960s, ethnicity-focused theories sympathized with civil rights. But for the ruling class to repudiate racial injustice, members of minority populations had to demonstrate their worthiness. As a result, by the late twentieth century, problems emerged:

> To treat race as a matter of ethnicity is to understand it in terms of culture. It is to undermine the significance of corporeal markets of identity and difference, and even to downplay questions of descent, kinship, and ancestry—the most fundamental demarcations in anthropology. Because cultural orientations are somewhat flexible—one can speak a different language, repudiate a previous religious adherence or convert to another, adopt a new "lifestyle," switch cuisine, learn new dances—ethnicity theories of race tend to regard racial status as more voluntary and consequently less imposed, less "ascribed."
>
> *(Omi and Winant, 2015)*

The implication is that the assignment of group identity according to physical appearance—the corporeal—exists as a tool for the powerful to oppress the powerless. It creates a system of domination and suppression. The distinctions between who is a citizen and who is not, who is a leader and who is not, and who is a member of the governing class and who is not, facilitate economic inequality, human subjugation, and imperial rule.

In the context of ethnicity-focused theories, individuals may view race as a cultural phenomenon, similar to other status-based identities: gendered groups (women's groups), groups of sexual orientation (lesbian, gay, bisexual, transgender, queer), or designated groups (Cuban Americans). Race, in other words, is an ethnic matter. Today, most of the paradigm's proponents have "moved rightward," within the framework of white racial nationalism rather than the civil rights movement, a consequence of an attempt to "understand race as a cultural phenomenon" (Omi and Winant, 2015).

Class

Class-focused theories of race assign differences to economic structures and processes. Affording primacy to economic relationships, the theories address distribution, exchange, and production. By joining race with class, however, the theories consider inequality. With inequality, the analysis includes society's unequal distribution of income and the exploitation of labor, both elements of capitalism, the prevailing economic system. With these two factors, a class system emerges.

Three approaches define different economic spheres: class conflict (Marxian), market relations (neoliberal), and systems of distribution (stratification).

Class conflict (Marxian) theory, in its classical form in the nineteenth and twentieth centuries, does not address race. But in the context of production, class conflict analyzes division, exploitation, and oppression of the working class. It, therefore, influences the class-focused paradigm of race. The idea begins with the social relations of production. Labor contributes to a value of output greater than its contribution as a resource input. The extra value flows to capitalists, owners of the means of production. With this arrangement, labor exploitation and inequality exist. Capitalists extract surplus value from laborers. With the desire to establish labor's share, racial division and discrimination grow (Omi and Winant, 2015).

The market relations (neoliberal) theory addresses market exchange and discrimination. First, discrimination results from the actions of the ruling class. Second, it stems from a systematic transfer of resources to the ruling class, who benefit from discriminatory practices. Third, it results from policies of the state, which acts on behalf of the ruling class. Together, these sources of discrimination rise in conjunction with capitalist systems. Race-focused labor laws, citizenship requirements, and exclusionism, implemented and sponsored by the state, perpetuate policies with racist outcomes. While the market relations theory attempts to reconcile inequity with market conditions, it struggles to achieve this result. Racial discrimination increases the cost of labor, so an efficient market should eliminate the practice. But the persistence of racial inequities results from the "extra-economic dimensions of racial formation: notably coercion and state action" (Omi and Winant, 2015).

Systems of distribution (stratification) theories address the reality that individuals with similar resources and economic opportunities exist in the same socioeconomic class. Within the ranks of hierarchy, social mobility occurs, shaping the maintenance and variation of social stratification. But during periods of rising inequality, individuals in lower socioeconomic classes struggle to rise into higher income quintiles. Connections, informal ties, recruitment, resource allocation, and social networks provide contours, either simplifying or complicating the process. Politics also play a role. The dynamics of authority and power reinforce the existing hierarchy. In this context, adaptation to new socioeconomic conditions, labor market opportunities, and cultural norms provide an opportunity for economic advancement. But inequity derives from both class-based stratification and race-based discrimination.

Nation

Nation-focused theories of race originate in the practice of empire. The conquest of territories in the modern world by European powers established colonies, the practice of nation-building, and imperial activities, resulting in nationalist movements that would eventually replace the empires. In modern times, the concept of the predominant socioeconomic class may ignore or dismiss the presence of people of color, their historical roles in the process of development, and their contributions in the evolution of beliefs, governing structures, and institutions. But the connection between nation and race has never been concrete. For the dominant class, it has been an imperative to improve or repair the concept of national identity. Over time, cycles of socioeconomic change demonstrate periods of racial politics, nationalism, and democratic initiatives. But histories of discrimination, exclusion, and immigration challenge the concept of foundational identity, leading to alternating periods of backlash and reform. In this context, race operates as a "multi-leveled organizing principle," establishing narratives for the existing order and linking the "visible characteristics of different social groups to different social statuses" (Omi and Winant, 2015). This practice establishes principles of exclusion and inclusion. The result is the forging of unity and solidarity on one hand and disunity and conflict on the other. As a result, while nationalism unifies the ruling class across differences in class, ethnicity, and status, it creates racial cleavages that give rise to the exploitation and exclusion of minority groups. As a contributing factor, state policy may either relax or tighten racial boundaries.

Assessment: the contrarieties of race

Race exists as a social construct, leading to oppression and resistance. This reality does not imply that race supersedes ethnicity or gender or exists as the most important element of intersectionality. Rather, race serves as a characteristic of inequity. For the purposes of domination and exploitation, racial hierarchies rank certain groups above others. As a social construct, racial identity perpetuates difference. This distinction shapes both conflicts and methods of appeasement, establishing race as an element of the existing order. Omi and Winant (2015) characterize the historical consequence:

> A great human sacrifice created the United States and all the Americas: the twin genocides of conquest and slavery. Although an immense effort has been made to repair the damage that sacrifice caused, the destruction can never really be undone. Much of the work of repair has been carried out by the victims themselves and their successors, who have tried to make a life on the gravesite of their ancestors and have sought to make "the destiny of America" finally theirs. That has not happened yet.

Racism permeates society

This discussion emphasizes the point that, while the Floyd murder during the coronavirus pandemic increased the awareness of racism and police misconduct, greater recognition of the problem did not alter power dynamics. Many individuals recognized the high cost to Black and Brown members of society from racial injustice, but policies with inequitable outcomes were embedded in historically grounded areas such as policing. Writing in the *New England Journal of Medicine*, Zinzi Bailey and her coauthors, Justin Feldman and Mary Bassett (2021), argue: "Confronting racism . . . requires not only changing individual attitudes, but also transforming and dismantling the policies and institutions that undergird the U.S. racial hierarchy." The reason is that racist policies reach back to the beginning of U.S. history. Ibram Kendi (2019) adds: "Racist ideas have defined our society since its beginnings and can feel so natural and obvious as to be banal, but antiracist ideas remain difficult to comprehend, in part because they go against the flow of this country's history." The framers of the U.S. Constitution did not solve the problem of racism. The Civil War did not end the legacy of racism. Over time, racism changed its form and perpetuated generational discrimination, existing in Jim Crow laws, residential segregation, and inequitable healthcare outcomes. As a result, individuals may be programmed to address human difference with apprehension and fear, handling difference by perpetuating it, ignoring it, or eliminating it, the latter serving as the most difficult option.

Modern societies have ethnic, linguistic, racial, and religious minorities, and therefore opportunities for racist policies and behavior. Violations of civil, cultural, economic, political, and social rights may take the form of discrimination or exclusion. In the United States, policies with racist outcomes exist for all minority groups, including African Americans, Asian Americans, Indigenous people, and members of the Latinx community. The reason for a comprehensive approach in dismantling racist policies is the durability of the concept, the perception of permanence. Because racist policies stretch across the centuries through attitudes, cultures, and institutions, actions to dismantle racist policies must involve all aspects of society, including the economy, education, health, and political system.

Moving beyond personal prejudices, implementing antiracist policies, and creating equitable outcomes requires intentionality. As an example, Bailey et al. (2021) argue that medical and public health communities may establish antiracist policies. The first responsibility is to document inequitable health outcomes. The second is to improve the availability of data on race and ethnicity. The third is to reflect on the medical establishment, focusing on racist practices and antiracist reforms. The fourth is to acknowledge that social movements challenge policies with racist outcomes. Together, the steps provide a framework to address inequities.

Antiracism

How to be an Antiracist, Ibram Kendi's (2019) book on antiracism as a transformative concept that "reorients and reenergizes" the discussion, argues that the function of racist ideas and bigotry is "to manipulate us into seeing people as the problem, instead of the policies that ensnare them." In this context, society trains individuals to see "deficiencies of people rather than policy." The implication,

according to Kendi, is that racist ideas lead people of color to "think less of themselves." But, in Kendi's framework, policies with racist outcomes maintain social inequities, including behavior, biology, culture, dueling consciousness, and
power. The way to undo racism, according to Kendi, is to identify, describe, and
dismantle it.

Behavior

With behavior, Kendi (2019) argues that the problem is to ascribe the actions
of an individual as indicative of the individual's group. But the behavior of individuals in any group, geographical circumstance, or cultural context reflects
the individual, not the group. A person who excels in school means the person is achieving academically, not that the person's group is inherently better
at education than another person's group. The individual should be praised for
being a good student, not a good student of a particular age, ethnicity, race, or
socioeconomic status. At the same time, if an individual struggles in school, the
problem does not demonstrate anything about the person's group. Personal behavior, in other words, does not imply a generalization about behavior within a
subset of the population. "Racial-group behavior is a figment of the racist's im-
agination. Individual behaviors can shape the success of individuals. But policies
determine the success of groups" (Kendi, 2019). In two ways, behavioral racism
impacts people's perception. First, behavioral racism implies that individuals are
responsible for the perceived behavior of racial groups. Second, behavioral racism makes racial groups responsible for the behavior of individuals. But to be an
antiracist is to "recognize that there is no such thing as racial behavior" (Kendi,
2019). Individual stories prove the behavior of individuals. "Just as race doesn't
exist biologically, race doesn't exist behaviorally" (Kendi, 2019). A framework of
antiracism means separating behavior from culture.

Biology

The survey of the entire human genome, completed in 2003, reveals that humans, regardless of race, are genetically more than 99.9 percent the same. An
important implication is our common humanity. But with human interaction,
instead of seeing an individual, we often see a race, an example of **racist categorizing**: identifying experiences with color-marked labels. With this view,
superficial differences connote different forms of humanity, the meaning of biological racism.

Racial distinction and hierarchy, when existing as widely held beliefs, perpetuate biological racism. This thinking includes two ideas: races are different in
their biology, and these differences create a system of hierarchy.

An antiracist, in contrast, identifies individuals as individuals. According to
Kendi (2019), "To be antiracist is to recognize the reality of biological equality,
that skin color is as meaningless to our underlying humanity as the clothes we

wear over that skin." To be antiracist is to identify the mirage of racial distinction and hierarchy, which elevates skin color above individuality.

But this reality does not imply a color-blind perspective. In our discourse, the elimination of racial categories discourages the identification of racial inequity. An inability to identify racial inequity means an inability to recognize racist policies. If we cannot recognize racist policies, it is not possible to dismantle them. "If we cannot challenge racist policies, then racist power's final solution will be achieved: a world of inequity none of us can see, let alone resist" (Kendi, 2019). Therefore, according to Kendi, to establish a world of antiracist policies, eliminating racist categories is not the first step but the last step.

Culture

Cultural attitudes, behaviors, and traits derive from both unique and common origins. A group may derive its culture from local surroundings and influences, historical conceptions of language, religion, interaction, and ideas from other groups. But whoever establishes the cultural standard creates cultural hierarchy. The attitudes, behaviors, and traits of one group may be viewed as superior to others. In Kendi's (2019) framework, "The act of making a cultural standard and hierarchy is what creates cultural racism. To be antiracist is to reject cultural standards and level cultural difference." Using "surface-sighted cultural eyes," Kendi argues, some in society may not identify or appreciate cultural forms, such as African cultural forms—especially customs, languages, and religions—believing they are overwhelmed by a dominant culture.

Kendi's point is that, if we refer to a group according to racial identity, such as Black Northerner or White Southerner, as opposed to Northerner or Southerner, we racialize the group. If we identify the culture of a group as lower on a cultural hierarchy, we establish it as inferior. Then every behavior and practice that does not correspond to the dominant culture is viewed as inferior. "Whoever creates the cultural standard usually puts themselves at the top of the hierarchy" (Kendi, 2019). But an antiracist sentiment considers groups and individuals with respect to their cultural history, influences, and practices, not an arbitrary standard of superiority.

Consciousness

Dueling consciousness means the simultaneous feeling of two identities or two strivings, such as feeling "Black" and "American." Dueling consciousness may lead to contradictions with a sense of belonging or place. According to Kendi (2019), to feel Black is to identify with a group, but to feel American is to strive to an existing ideal, which may not correspond to the group's identity. Dueling consciousness leads to a conflict between assimilationists, **segregationists**, and antiracists. Assimilationists express "the racist idea that a racial group is culturally or behaviorally inferior and is supporting cultural or behavioral enrichment programs to develop that racial group" (Kendi, 2019). Segregation-ists express "the racist idea that a permanently inferior racial group can never be developed and is supporting policy that segregates away that racial group" (Kendi, 2019). With these concepts, a culturally dominant group may experi-ence its own dueling consciousness. While assimilationist ideas suggest a group is "temporarily" inferior, segregationist ideas suggest a group is "permanently" inferior. The sentiments may, therefore, conflict. But both are racist ideas. In contrast, antiracists express "the idea that racial groups are equals and none needs developing" (Kendi, 2019).

Power

Kendi (2019) argues that the social construct of race "creates new forms of power: the power to categorize and judge, elevate and downgrade, include and exclude. Race makers use that power to process distinct individuals, ethnici-ties, and nationalities into monolithic races." The implication is that members of the ruling class may batter individuals from other groups for their perceived differences, including class, disability, ethnicity, gender, religion, or sexuality. In this framework, the ruling class informs the notion of identity and confers privilege, the main privilege being a person who is legal, normal, and stand-ard. Race as a social construct means making racial hierarchies. With hierar-chies, those who benefit the most from the existing order are characterized with positive qualities—inventive, smart, vigorous—but those who benefit the least are characterized with negative qualities—greedy, opinionated, strict. Racism normalizes and rationalizes these differences, thus motivating, maintaining, and propagating social inequities. "This cause and effect—a racist power creates racist policies out of raw self-interest; the racist policies necessitate racist ideas to justify them—lingers over the life of racism" (Kendi, 2019). The result is a power dy-namic that maintains the existing order.

Intervention and reform

Antiracism intervention means the "action-oriented, educational and/or po-litical strategy for systemic and political change that addresses issues of racism and interlocking systems of social oppression" (Calliste and Dei, 2000). Exam-ples include antidiscrimination legislation, organizational change, and equita-ble policy reforms. This definition acknowledges both the multiple contexts of racial inequity and different forms of intervention (Calliste and Dei, 2000). For marginalized communities, interventions address institutionalized racism (inequitable access to opportunities and output), personally mediated racism (discrimination and prejudice), and internalized racism (negative actions, atti-tudes, and beliefs) (Hassen et al., 2021). Antiracism interventions, identified in studies on healthcare but applicable to police brutality, residential segregation, and environmental racism, include individual-level interventions (antiracist

TABLE 5.1 Principles of antiracism intervention

Number	Principle
1	Define the problem and set clear goals and objectives
2	Incorporate explicit and shared antiracist language
3	Establish leadership buy-in and commitment
4	Invest dedicated funding and resources
5	Include the right expertise and support
6	Establish meaningful community and individual partnerships

Source: Hassen et al. (2021).

training and critical reflection of attitudes, beliefs, knowledge, and practices), community-level interventions (development of partnerships between groups, engagement in the decision-making process, and reorganization of power structures), organizational-level interventions (antiracism policy, collection of data on policy outcomes, and commitments to structural change), and policy-level interventions (accountability, inclusive participation, monitoring, recruitment and retention of people of color, and transparency) (Hassen et al., 2021).

Principles of antiracism intervention, according to Hassen et al. (2021), establish a framework of reform (Table 5.1). First, policy practitioners and invested citizens should define the problem and set clear goals and objectives. A successful first step aligns interventions with intended goals, increasing the potential to achieve desired outcomes. Second, policy practitioners and invested citizens should incorporate explicit and shared antiracism language. A successful second step leads to an agreement on the meaning of diversity, inclusion, and cultural awareness. Third, policy practitioners and invested citizens should establish leadership buy-in and commitment. At different levels of organization, the meaningful involvement of leaders and institutions facilitates effective processes. Fourth, policy practitioners and invested citizens should invest dedicated funding and resources. This principle ensures the means with which to accomplish policy goals. Fifth, policy practitioners and invested citizens should include the right expertise and support. Because it is challenging to implement policy with equitable outcomes, capable personnel must participate in the process. Sixth, policy practitioners and invested citizens should establish meaningful and ongoing partnerships. This principle ensures an appropriate framework. Together, antiracism interventions dismantle racist policies, including those that lead to police brutality, residential segregation, environmental racism, and health inequities.

Police brutality

Of all the countries in the world, the United States has the highest rate of incarceration. The country's police kill civilians at a higher rate than police in other wealthy countries. "The history of courts, prisons, and police as institutions that maintain racial hierarchy is key to understanding the deeply punitive and racially unequal nature of the U.S. criminal legal system" (Bailey et al., 2021). In the criminal legal system, racial bias and inequitable outcomes result from police encounters and the length of sentencing. In the United States, contemporary policing has its roots in slavery. Slave patrols, established in Virginia in the eighteenth century, captured runaway slaves and quelled uprisings. After the abolition of slavery and attempts at reform during the Reconstruction, the criminal justice system, including police and prisons, asserted dominance for the ruling class. In many contexts, law enforcement "sanctioned, enabled, and participated in" the subjugation of minority members of the population (Bailey et al., 2021). In the United States, the War on Crime in the 1960s and the War on Drugs in the 1970s portended an increase in the rate of incarceration. The trend continues today.

An important question persists: what factors create an environment that is conducive for police brutality, exemplified by the murders of Breonna Taylor, George Floyd, and many others? It is a difficult question to answer. Not all members of law enforcement exhibit behaviors that lead to police brutality. But "policing has long been entangled in other structures that reproduce racism" (Bailey et al., 2021). Because racial injustice persists in the law enforcement community, often with violent consequences, the idea of "police reform" remains incomplete. The historical lens reveals racial subjugation. For successful intervention to occur, society must identify the sectors (economy, housing, public health) that require an influx of resources, reallocated from policing, that do not require a response from the institution of law enforcement.

Activist movement

After the George Floyd murder, on May 25, 2020, the social activist movement **Black Lives Matter** (BLM) increased in energy, enthusiasm, and organization. BLM, which began as a protest for the killing of a Black man, Travon Martin, in 2012, escalated during the coronavirus pandemic, in 2020. The 2020 protests started in Minneapolis with memorials in what is now known as George Floyd Square: the four-block area in a mixed-income neighborhood in South Minneapolis that serves as a shrine with candles, flowers, and messages. The square is populated by activists, community members, mourners, protestors, and tourists. Over time, the area became "more than a shrine to Floyd's life; it was a monument to others who had died in encounters with police, and a headquarters for an emergent movement" (Cobb, 2021). In a nearby field, community members placed dozens of symbolic gravestones.

Throughout the summer of 2020, the BLM movement spread throughout the nation, calling for freedom, justice, and healing, "demanding a societal reckoning with the racist foundations of this country (United States) and the ongoing structural violence that limits the life changes of people of color" (Crooks et al., 2021). But protests occurred on a global scale, in cities such as Barcelona, Lagos,

London, Paris, Sydney, and hundreds of others. In the United States, on June 6, 2020, half a million people demonstrated in support of BLM. As enthusiasm grew, by the beginning of July 2020, upwards of 26 million people in the United States, 10 percent of the adult population, participated in the demonstrations, "the largest movement in the country's history" (Buchanan et al., 2020). The BLM movement intended to end police brutality, eradicate white supremacy, establish equity for marginalized groups, and build power for communities to intervene in the case of police violence.

The BLM movement resulted from both the George Floyd tragedy and the pent-up anger about other cases of police brutality. But anxiety from pandemic lockdowns, frustration about the actions of the sitting president (Donald J. Trump), insecurity about the future trajectory of the novel coronavirus, and rising levels of unemployment also served as important factors. BLM helped to define the era of cascading crises, served as a racial reckoning, and ushered in new attitudes toward race and justice. It also led to policy reform. One year after the Floyd murder, "more than 30 states . . . passed more than 140 police oversight and reform laws" (Chudy and Jefferson, 2021). As a result, BLM served as an influential and visible movement.

The police officer Derek Chauvin, who knelt on Floyd's neck for 9 minutes and 29 seconds while three officers stood by until Floyd died, was convicted on April 20, 2021 with two counts of murder and one count of manslaughter. He was sentenced to 22½ years in prison. On the one hand, many people viewed the verdict as a "just resolution to a public tragedy" (Cobb, 2021). The reopening of George Floyd Square to traffic served as part of a spirit of relief and a desire to move on from the experiences of police brutality. The tensions in Minneapolis had served as a microcosm of the national debate about injustice, race, and policing. On the other hand, held with equal resolve, the Chauvin trial existed as one element of many that needed to be addressed before experiencing reform. Many saw Floyd's death as a "singular incident of spectacular violence," but those involved in the web of police brutality were "more likely to connect his death to a long genealogy of events that both preceded and followed it" (Cobb, 2021). Less than a month after testimony began in the trial, American law enforcement had killed at least 64 other people, with Black and Latinx individuals constituting more than half of the deaths (Smith, 2021).

By the end of 2021, support for BLM declined. According to Jennifer Chudy and Hakeem Jefferson (2021), three reasons existed. First, the surge reflected "shock and disapproval over this particular episode (the George Floyd murder) rather than a broad embrace of a political movement." The video of the murder prevented individuals from establishing their own narratives about police brutality. Second, the event occurred during the coronavirus pandemic, which provided an attentive and anxious audience. A large number of people were moved by the murder of a helpless man. Third, a decline in support resulted from the "increased politicization of the issue by elites." Rather than focusing on police injustice, many who opposed reform turned their attention to the protestors and the costs of social upheaval (Chudy and Jefferson, 2021).

Intervention and reform: the case of police brutality

When it comes to enforcing the law and maintaining public order, society should hold police officers to the highest standard. But in law enforcement, police brutality persists. It is, therefore, important to assess the culture that allows this behavior to "burrow deeply into a department's ranks" (Grant, 2021). To begin a process of reform, police officers who commit illegal acts should be fired. For all other officers, transparent and accessible records of discipline should exist. Police unions should not maintain officers on staff with records of misconduct and racism. For all members of society, the system of law enforcement should ensure safety and opportunity, not enforce second-class status for vulnerable members of the population. Training on the intersection of race, class, gender, socioeconomic status, and violence should accompany employment status. Education on conflict management and de-escalation should continue.

In the current era, police officers are examples of the overloaded state, asked to address problems outside of crime fighting, such as ensuring public safety, guaranteeing the sanctity of the judicial system, and protecting property. Additional responsibilities include areas in which other members of the professional class are better suited: domestic violence, family breakdown, juvenile delinquency, and mental health. In this context, the institution of policing is ripe for reform. The call to defund the police by eliminating police departments, a rallying cry after the George Floyd murder, exists as a first option for reform in a process to deconstruct the police. To deconstruct the police means shifting resources from police forces to more appropriate professions, including counselors, psychologists, and social workers, who address social inequities but not crime fighting (Micklethwait and Wooldridge, 2020).

FIGURE 5.1 Forceful police actions.
Source: Human Health Watch (2019).

Fighting poverty constitutes a second opportunity for reform. While a 2019 Human Rights Watch Report identified strong evidence of racial bias in police activity, much of the disparity with respect to forceful action relates to "concentrated policing in high poverty neighborhoods, which are more frequently communities of color" (Figure 5.1). When compared with the majority class, minority populations experience inferior educational opportunities, health outcomes, social status, early-childhood education, school systems, and medical care. Public policies may alleviate these problems, such as the extended child tax credit, universal basic income, and universal healthcare, but the systematic nature of the racism often remains. In underserved neighborhoods, better access to education, employment, and healthcare leads to greater lifetime opportunities. Universal preschool, paid maternity/paternity leave, and childcare reduce poverty. A public sector that helps the poor and needy as much as the rich levels the economic playing field, reduces crime, and complements the process of police deconstruction.

Residential segregation

In Chicago, while the coronavirus pandemic raged, a group of activists engaged in a hunger strike to stop the relocation of a highly polluting metal recycling plant to a working-class and industrial neighborhood. In urban areas, poorer neighborhoods often experience the external costs of manufacturing, including pollution, congestion, and negative health effects. The clustering of industrial activity in poorer neighborhoods results from a political reality: residents of more prosperous neighborhoods influence lawmakers to implement favorable zoning laws. In the case of Chicago, the management company of General Iron, a century-old metal recycling plant, wanted to move the facility from a northside and industrial location to a poorer southside neighborhood. This choice irked many southside residents, leading to the hunger strike. The General Iron plant, while providing jobs, had a history of violations, including citations for nuisance and pollution. While the company had a chance to update their production process, modernize technology, and reduce costs, projected increases in air pollution and asthma-related deaths on the south side would increase the neighborhood's level of environmental toxicity. Even though this case occurred in Chicago, industrial clustering often exists in or near poorer neighborhoods.

Redlining

The United States has a history of residential segregation. In 1933, the federal government established the Homeowners' Loan Corporation (HOLC), expanding home ownership during the Great Depression. But the HOLC used maps of more than 200 cities with the racial composition of neighborhoods. To clarify the lending process, the HOLC drew red lines around African American neighborhoods. What was the result? "Redlining made mortgages less accessible, rendering prospective Black homebuyers vulnerable to predatory terms, thereby increasing lender profits, reducing access to home ownership, and depriving these communities of an asset that is central to intergenerational wealth transfer" (Bailey et al., 2021). The government-sanctioned practice of redlining created the model of segregated neighborhoods, which persisted in Chicago and other cities, including Detroit, Los Angeles, and St. Louis. Redlining validated racist policies, such as locating manufacturing in or near minority neighborhoods, restricting housing covenants, and undervaluing real estate. Although the Fair Housing Act of 1968 eliminated redlining, residential segregation remained, with decades of disinvestment in poorer neighborhoods.

Segregation and the coronavirus

During the pandemic, minority members of the population had higher rates of Covid-19. In poorer neighborhoods, the virus surged during multiple infection waves. But members of high-income households worked from home or fled to rural settings, options unavailable to their lower-income counterparts. In poorer communities, the higher rates of morbidity and mortality resulted from less access to healthcare, pre-existing medical conditions, and lower rates of vaccination. Other factors included residential segregation and income inequality, which correlated with poverty. During the pandemic, not only did these variables link to the racial composition of neighborhoods, but they also led to higher rates of infection and death. Residents of high-poverty, low-resource, and population-dense neighborhoods experienced limited access to healthcare, more pressure to maintain frontline jobs, and fewer vaccination centers. As the pandemic surged, lower-income neighborhoods had fewer resources to break the cycle of infection. This reality enforced the roots of inequality, including discrimination, less economic opportunity, and residential segregation.

Intervention and reform: the case of residential segregation

Residential segregation separates individuals by race and/or ethnicity, restricts community resources, and leaves households with less access to the resources of modern society. These neighborhoods experience less public- and private-sector investment, lower rates of economic growth, and less upward mobility. In contrast, wealthier neighborhoods have more access to education, employment, and public health. To address the problem of racial inequity, society should alter the conditions of residential segregation. Successful reforms include, but are not limited to, the creation and promotion of affordable housing in gentrifying areas, direct investment for education, infrastructure, and job creation, equitable zoning policies, incentives for density near public transportation, and programs of antidiscrimination.

Environmental racism

Policies with inequitable outcomes may increase exposure to environmental pollution: "Social scientists and epidemiologists have long understood that racism is a fundamental cause of disease that operates through complex, ever-changing mechanisms" (Nigra, 2020). The health effects of racial inequity are so well-pronounced that in 2015 the American Public Health Association began a campaign against racism, recommending an antiracist agenda. The agenda identified the link between health policy and inequitable outcomes, engaged in a campaign to protect the most vulnerable members of the population, and anticipated how racism evolved to create new disparities in environmental exposure (Nigra, 2020).

Both historical and contemporary policies that perpetuate environmental pollution, including municipal zoning and urban development, correlate with health disparities. In many disadvantaged neighborhoods, the proximity to environmental hazards and a lack of public health services prolong health inequities. Among people of color, examples include relatively higher levels of lead in the blood, greater exposure to air pollution, and closer proximity to toxic chemical emissions. These problems situate the inadequacy of both public health services and programs of environmental protection within the context of environmental racism.

Consider air pollution. In quality-of-life calculations, clean air correlates with healthy lifestyles. But clean air is not evenly distributed. People of color are more likely to live near polluting factories: "political, socioeconomic, and discriminatory forces have concentrated people of color within distinct and often socioeconomically disadvantaged neighborhoods in which they are also disproportionately exposed to environmental hazards, including ambient air pollution" (Woo et al., 2019). Through respiratory and cardiovascular diseases, inferior air quality links to morbidity and mortality. Lacking economic resources and social capital, individuals in these neighborhoods struggle to prevent the location of polluting factories. In Chicago, the General Iron case serves as an example. When neighborhood infrastructure diminishes, property values decline, and residents lose their ability to relocate, inferior air quality may persist (Woo et al., 2019).

Environmental racism and the coronavirus pandemic

During the coronavirus pandemic, toxic living environments inflated Covid-19 death rates. Crowded housing conditions contributed to the disparity in health outcomes. With Covid-19, both poverty and environmental racism were linked to morbidity and mortality; however, Harriet Washington (2020), writing in *Nature*, distinguished between the two, using the United States as context:

> Racial disparities in exposure to environmental pollutants are greater factors that remain even after controlling for income. African Americans who earn US$50,000–60,000 annually—solidly middle class—are exposed to

much higher levels of industrial chemicals, air pollution and poisonous heavy metals, as well as pathogens, than are profoundly poor white people with annual incomes of $10,000. The disparity exists across both urban and rural areas.

Exposure to environmental hazards led to lower life expectancies and pandemic inequities. Those suffering from pollution also suffered from inferior health outcomes from the virus. Risk factors for Covid-19 existed "within the context of environmental contaminants and the adverse social determinants of health that put minority communities at increased risk for disease and mortality" (Njoku, 2021). The social determinants of health included education and training, employment and income, health and medical care, and the physical environment. Environmental racism impacts each social determinant, by weakening opportunities for education and training, creating inferior economic opportunities and health outcomes, and polluting the physical environment.

Case study 5.1 Line 3, oil, and threats to an Indigenous community

During the coronavirus pandemic, a policy that led to racist outcomes occurred in northern Minnesota, in a fight between an Indigenous community, the Ojibwe, and Enbridge Inc., the Canadian multinational oil company. Controversy existed over an Enbridge project, Line 3, a pipeline transporting tar sands oil, the dirtiest in the marketplace according to carbon content, over a 1,700-kilometer (1,000-mile) route from Edmonton, Alberta to Superior, Wisconsin. Enbridge had a history of oil spills. In addition, the pipeline traversed both the head waters of the Mississippi River and wild rice fields. As a result, activists, organizers, and protestors among the Ojibwe and their allies fought pipeline construction. According to the opposition, the pipeline infringed on treaty rights, threatened local water supplies, and contributed to climate change. The case related to the coronavirus pandemic, because the virus spread through "man camps" along the pipeline route, the temporary residences of out-of-state pipeline workers. According to Winona LaDuke, Native rights activist and member of the Ojibwe tribe, "They're cutting, they're grinding, they're welding, they're smashing, they're laying pipe. They're all around you, and they're coming toward you. That's pretty traumatic. A lot of cops, a lot of destructive equipment, a lot of people scared" (Marchese, 2021). A policy of pipeline construction led to inequitable outcomes for the Indigenous community, including the destruction of natural resources. But it possessed few benefits, except the economic gains for Enbridge shareholders and funds for local police. The proposed flow of oil through the pipeline, with an initial capacity of 760,000 barrels per day, represented less than 1 percent of global production. Even more, a gallon of gasoline made from tar sands oil led to 15 percent more carbon dioxide emissions than a gallon of gasoline made from conventional oil. Because the pipeline violated Indigenous treaty rights—signed by the U.S. federal government—the case served as an example of environmental racism.

Intervention and reform: the case of environmental racism

Because environmental racism exists in a historical context, society should establish a comprehensive framework to solve the problem. Harriet Washington (2020) provides a perspective:

> We need to take a longer, harder looks at environmental racism—systems that produce and perpetuate inequalities in exposure to environmental pollutants. These can persist even in the absence of malevolent actors. The main culprits include indifference and ignorance, inadequate testing of industrial chemicals, racism, housing discrimination, corporate greed and lax legislation.

First, society should collect data on environmental racism. What behaviors establish segregated neighborhoods? When negative outcomes exist, who bears the burden? Are the victims people of color? What zoning laws perpetuate the problem? Second, after identifying and recognizing the problem, policymakers should implement reforms that break the link between environmental hazards and inequitable outcomes. Additional questions exist. In urban areas, how should society reduce pollution flows? In rural areas, how should society prohibit the desecration of natural resources? Third, society should recognize that the variables that correlate with environmental racism, including lower levels of educational attainment, less access to healthcare, and less income, are the same variables that constitute lower socioeconomic status. As the head of the World Health Organization, Tedros Adhanom Ghebreyesus, argues, "No one is safe until everyone is safe" (Washington, 2020). To overcome environmental racism, society must identify it, confront it, and implement equitable solutions.

Health inequities

The SARS-Cov-2 virus was dangerous because of its high level of transmissibility. During multiple infection waves, hospitals struggled to cope. Medical shortages and a lack of hospital beds, nurses, and ventilators, the ability of the virus to mutate, and the inability of many individuals to seek vaccinations perpetuated the crisis. But inequitable outcomes also related to social conditions. In the United States, "African Americans, Hispanic/Latinx persons, and American Indian or Alaska Native, Non-Hispanic persons (were) more likely to contract, be hospitalized with, and die from Covid-19, when compared to non-Hispanic whites" (Njoku, 2021). That is, racialized conceptions of the pandemic permeated the medical landscape. Consider a historical context. A 2018 report argued that "Black, American Indian and Alaska Native, and Native Hawaiian and Pacific Islander patients continued to receive poorer care than White patients . . . with little to no improvement from decades past" (Agency for Healthcare Research and Quality, 2018). Health inequities persist.

Reasons for health inequities

Health inequities relate to socioeconomic conditions and flexible resources (Figure 5.2). With socioeconomic conditions, racism serves as a propagating factor, contributing to inequities in income and education. With access to flexible resources, the topic of this section, several factors exist, according to Bailey et al. (2021) and Phelan and Link (2015). First, in healthcare systems, discrimination and racial beliefs contribute to inequitable outcomes among patients. Second, when vulnerable members of the population have pre-existing conditions, healthcare inequities persist. Third, neighborhoods create social connections, but racial segregation reduces access to healthcare resources. Fourth, neighborhood disinvestment complicates the recruitment of clinical care workers. Fifth, discrimination, inferior life circumstances, and poor health conditions lead to physiological and psychological stress. Together, the "actions by parties ranging from medical schools to providers, insurers, health systems, legislators, and employers have ensured that racially segregated . . . communities have limited and substandard care" (Bailey et al., 2021).

For vulnerable individuals, health inequities are "pronounced, persistent, and pervasive" (Sondik et al., 2010). While racism exists on an individual level—such as doctor to patient—systematic inequities perpetuate and exacerbate the problem, leading to inferior outcomes for those with higher rates of cancer, heart

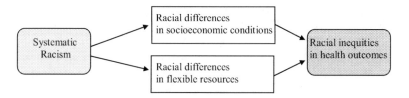

FIGURE 5.2 Racism as a cause of health inequities.
Source: Adapted from Bailey et al. (2021) and Phelan and Link (2015).

disease, and infant mortality. One reason is that marginalized members of the population benefit less from medical advances, the systems translating and using new technologies. A second reason is that racial biases in both healthcare and treatment lead to doubt and uncertainty with vaccinations. A third reason is that ideologies, institutions, and social forces interact to reinforce health inequities. A fourth reason is that structural mechanisms reconstitute the "conditions necessary to ensure their perpetuation" (Gee and Ford, 2011). A fifth reason is that poor doctor communication may lead to delayed or foregone care (Rhee et al., 2019). With healthcare, these factors perpetuate policies with racist outcomes.

Intervention and reform: the case of health inequities

Inequitable health outcomes are rooted in and supported by the mistaken belief that racial and/or ethnic groups are "intrinsically disease-prone and, implicitly or explicitly, not deserving of high-quality care" (Bailey et al., 2021). As with police brutality, residential segregation, and environmental racism, health inequity does not exist in isolation. Rather, daily practices reify race, contributing to the mistaken belief of race as an intrinsic biological difference. Even more, with public health, the concept of **racialization**—when groups are socially constructed as unequal and different because of race—intersects with other characteristics of identity, including ageism, poverty, sexism, and xenophobia. Gee and Ford (2015) recommend measuring racism in period-specific ways, developing standardized approaches for estimating intergenerational effects, and expanding support for resources. The argument for intervention, therefore, addresses health inequities, historical context, and intersectionality, establishing a comprehensive platform to replace inequitable outcomes with equitable alternatives.

Summary

A contrast exists between antiracist and racist sentiments. Through their actions, antiracists support antiracist policies and ideas. Antiracist polices lead to equitable outcomes for all groups. Antiracist ideas mean that racial groups are equal in their apparent differences. But racists support racist polices and ideas. Racist policies lead to inequitable outcomes for specific groups. Racist ideas mean that racial groups are unequal in their apparent differences. In a historical context, racist policies and ideas exist as part of structured and well-established systems. Policies and ideas with racist outcomes maintain social inequities, including the areas of behavior, biology, culture, dueling consciousness, and power. Examples of police brutality, residential segregation, environmental racism, and health inequities demonstrate that dismantling racism requires a comprehensive approach. Antiracism interventions dismantle racist policies and ideas.

Key terms

Antiracism intervention

Antiracist

Black Lives Matter

Dueling consciousness

Institutions

Racist

Racialization

Racist categorizing

Segregationists

Chapter takeaways

LO1   During the coronavirus pandemic, problems of racial injustice persisted.

LO2   Systems that facilitate inequality and privilege include several factors, not just race, that intersect to create bias, discrimination, and stratification.

LO3   The way to undo racism is to identify, describe, and dismantle it.

LO4   Principles of antiracism intervention establish a framework for reform.

LO5   Policing has long been entangled in other structures that reproduce racism.

LO6   During the coronavirus pandemic, residents of high-poverty, low-resource, and segregated neighborhoods experienced limited access to healthcare, more pressure to maintain frontline jobs, and fewer vaccination centers.

LO7   In many disadvantaged neighborhoods, the proximity to environmental hazards and a lack of public health services prolong health inequities.

LO8   Health inequities relate to differing socioeconomic conditions, access to flexible healthcare resources, and racism.

Questions

1   Why did the George Floyd murder create a lasting legacy?

2   With respect to Kendi's (2019) framework, contrast antiracist and racist sentiments. For policy implementation, how do the two concepts differ?

3   In many modern contexts, racist policies permeate society. Why?

4   With respect to law enforcement, residential segregation, environmental racism, or public health, how do systems of privilege and inequality perpetuate inequitable outcomes? How does the coronavirus pandemic influence the framework of analysis?

5   What are the characteristics of antiracist behavior? Concerning policy, identify and discuss an example.

6   With respect to a particular form of policy inequity, how do the foundational principles of antiracism intervention apply?

7   With respect to police brutality or residential segregation, how should society dismantle racist outcomes? With respect to your proposed changes, be specific.

8   With respect to environmental racism or health inequities, how should society dismantle racist outcomes? With respect to your proposed changes, be specific.

References

Agency for Healthcare Research and Quality. 2018. *National Healthcare Quality and Disparities Report.* Rockville, MD: Agency for Healthcare Research and Quality.

Bailey, Zinzi, Feldman, Justin, and Bassett, Mary. 2021. "How structural racism works—racist policies as a root cause of U.S. racial health inequities." *New England Journal of Medicine,* 384(8): 768–773.

Buchanan, Larry, Bui, Quoctrung, and Patel, Jugal. 2020. "Black Lives Matter May Be the Largest Movement in U.S. History," *The New York Times,* July 3.

Calliste, A. and Dei, G. 2000. *Power, Knowledge and Anti-Racism Education: A Critical Reader.* Halifax: Fernwood Publishing Co.

Choi, Jung, McCargo, Alanna, Neal, Michael, Goodman, Laurie, and Young, Caitlin. 2019. "Explaining the Black-White Homeownership Gap: A Closer Look across Disparities in Local Markets." *Urban Institute,* October.

Chudy, Jennifer and Jefferson, Hakeem. 2021. "A Moment, Not a Reckoning." *The New York Times,* May 23.

Cobb, Jelani. 2021. "The Free State of George Floyd." *The New Yorker,* July 12 & 19.

Crooks, Natasha, Donenberg, Geri, and Matthews, Alicia. 2021. "Ethics of research and the intersection of Covid-19 and black lives matter: a call to action." *Journal of Medical Ethics,* 47: 205–207.

Galea, Sandro and Abdalla, Salma. 2020. "Covid-19 pandemic, unemployment, and civil unrest." *JAMA,* 3254(3): 227–228.

Gee, Gilbert and Ford, Chandra. 2011. "Structural racism and health inequities: old issues, new directions." *Du Bois Review,* 8(1): 115–132.

Grant, David. 2021. "We Sweep Uncomfortable Issues Under the Rug." *The New York Times,* April 18.

Hassen, Nadha, Lofters, Aisha, Michael, Sinit, Mall, Amita, Pinto, Andrew, and Rackal, Julia. 2021. "Implementing anti-racism interventions in healthcare settings: a scoping review." *International Journal of Environmental Resources and Public Health,* 18, 2993.

Human Rights Watch. 2019. *Get on the Ground: Policing, Poverty and Racial Inequality in Tulsa, Oklahoma.* New York: Human Rights Watch.

Kendi, Ibram. 2019. *How to Be an Antiracist.* London: One World.

Marchese, David. 2021. "Winona LaDuke Feels That President Biden Has Betrayed Native Americans." *The New York Times Magazine,* August 6.

Micklethwait, John and Wooldridge, Adrian. 2020. *The Wake-Up Call: Why the Pandemic Has Exposed the Weakness of the West, and How to Fix It.* New York: HarperVia.

Nigra, Anne. 2020. "Environmental racism and the need for private well protections." *PNAS,* 117(30): 17476–17478.

Njoku, Anuli. 2021. "Covid-19 and environmental racism: challenges and recommendations." *European Journal of Environment and Public Health,* 5(2): em0079.

Omi, Michael and Winant, Howard. 2015. *Racial Formation in the United States.* New York: Routledge.

Phelan, Jo and Link, Bruce. 2015. "Is racism a fundamental cause of inequalities in health?" *The Annual Review of Sociology,* 41: 311–330.

Rhee, Taeho, Marottoli, Richard, Van Ness, Peter, and Levy, Becca. 2019. "Impact of perceived racism on healthcare access among older minority adults." *American Journal of Preventive Medicine,* 56(4): 580–585.

Shaw, Susan and Lee, Janet. 2012. *Women's Voices Feminist Visions: Classic and Contemporary Readings.* New York: McGraw-Hill.

Smith, Talmon. 2021. "Whose Racial Reckoning Was It?" *The New York Times,* May 23.

Sondik, Edward, Huang, David, Klein, Richard, and Stacher, David. 2010. "Progress toward the healthy people: 2010 goals and objectives." *Annual Review of Public Health*, 31(1): 271–281.

Washington, Harriet. 2020. "How environmental racism fuels pandemics." *Nature*, 581: 241.

Woo, Bongki, Kravitz-Wortz, Nicole, Sass, Victoria, Crowder, Kyle, Teixeira, Samantha, and Takeuchi, Davis. 2019. "Residential segregation and racial-ethnic disparities in ambient air pollution." *Race and Social Problems*, 11(1): 60–67.

6

DOMESTIC AND FAMILY VIOLENCE

Chapter learning objectives

After reading this chapter, you will be able to:

LO1   Explain the shadow pandemic of domestic and family violence.
LO2   Provide evidence of domestic and family violence during the coronavirus pandemic.
LO3   Recognize the universality of the problem.
LO4   Address the diverse contexts in which the problem exists.
LO5   Evaluate theories of domestic and family violence.
LO6   Discuss programs of intervention and methods of policy reform.
LO7   Consider lessons of the shadow pandemic.

Chapter outline

Shadow pandemic
Evidence
Universality
Diverse contexts
Theories
Intervention and policy reform
Lessons
Summary

Shadow pandemic

During the coronavirus pandemic, the United Nations identified the rise in domestic and family violence (DFV) as a **shadow pandemic**, occurring on a

DOI: 10.4324/9781003310075-8

global scale (Mlambo-Ngcuka, 2020). The evidence across countries demonstrated that DFV acted like an "opportunistic infection, flourishing in the conditions created by the pandemic" (Taub, 2020). By increasing anxiety, isolation, and stress, lockdown restrictions exacerbated the problem. During this time, "governments largely failed to prepare for the way the new public health measures would create opportunities for abusers to terrorize their victims" (Taub, 2020). During lockdown, victims could not go outside, seek help, or rely on social services. Organizations that were supposed to protect victims, already underfunded, buckled from a shortage of resources. As the pandemic progressed, many governments scrambled to rectify the mistake, but they could not prevent a rise in DFV.

The pattern

Across countries, a pattern existed. When governments implemented lockdown measures without consideration of external effects, the unintended side effects of policy implementation, DFV increased. Because of the risk of infection, shelters struggled to take victims. Extended family and friends could not intervene. There was no place to hide. Only when the problem became clear did some governments respond, publicizing hotlines and apps, mobilizing civic groups and social services, and advising frontline workers in healthcare, when possible, to assist. Some communities even declared that, if victims needed refuge, they could ignore lockdown measures.

The definition

For context, DFV is defined to include behaviors that intend to control, intimidate, or manipulate a family member, former family member, or partner. Behaviors include economic abuse, emotional abuse, harassment or stalking, isolation, physical assault, psychological abuse, sexual assault and abuse, social abuse, verbal abuse, and threats that coerce victims into acceptance and compliance. The distinguishing feature of DFV, control, occurs when perpetrators feel they are "entitled" to abuse others, and when that behavior is supported by both cultural norms and social expectations (Wendt and Zannettino, 2015). A broad definition helps to ensure that "any harmful behavior, no matter how major or minor it may seem to the victim, perpetrator, or people outside the interpersonal relationship, is captured and understood in a way that facilitates a relevant service response" (Meyer and Frost, 2019). On a global scale, DFV has been recognized as a violation of human rights: "The World Health Organization argues that violence against women is the most pervasive yet under-recognized human rights violation in the world" (Ellsberg and Heise, 2005). Jennifer Nixon and Cathy Humphreys (2010) state:

> Of all the claims about domestic violence . . . the assertion that intimate
> partner violence is an established and widespread feature of both OECD

(Organization for Economic Cooperation and Development) and developing countries is perhaps the most important. The framing of domestic violence as a common occurrence has been consistently communicated by the movement (against domestic violence), from the earliest days of conscious-raising groups to the work of contemporary service-provision organizations and is supported by the survey data from North America, Australia, and the UK.

Calls to eliminate the problem by women's rights and health organizations have experienced limited success. According to a report by the World Health Organization (WHO, 2021), one in three women around the world reports physical or sexual violence, normally by a domestic partner. In every country and culture, the report emphasizes, violence is endemic, regularly found in the population: 10 percent of ever-married or partnered women around the world "have been subjected to physical and/or sexual intimate partner violence at some point within the past 12 months" (WHO, 2021).

DFV and crises

Although debate exists whether large-scale crises such as pandemics create pathways to DFV, the variables correlate. It is clear, however, that these emergencies increase household anxiety, stress, and uncertainty. Large-scale crises break down economic and social infrastructures, including family and support groups, employment networks, and neighborhood oversight, compounding the problem (Kofman and Garfin, 2020; Rubenstein et al., 2020). After the 2004 Indian Ocean earthquake and tsunami, DFV increased (MacDonald, 2005). In the 2009 aftermath of Hurricane Katrina, a four-fold increase in DFV occurred among displaced women in Mississippi (Anastario et al., 2009). Studies also found a rise in DFV after earthquakes, floods, and hurricanes (Parkinson and Zara, 2013).

DFV and the coronavirus pandemic

The coronavirus pandemic served as a unique case. In order to slow the spread of the virus, governments implemented unprecedented forms of control. But quarantines, economic shutdowns, and school closings led to isolation, rising unemployment, and remote learning. On a macro scale, the movement from organization and routine to disorganization and uncertainty stretched the limits of institutional support, the competence of leaders, and economic assistance. For countries, an inability to forecast both the spread of disease and economic downturn limited the effectiveness of policy intervention. On a micro scale, lockdown measures limited the ability of households to maintain routines. When perpetrators inflicted situational forms of abuse, violence existed as a method to control an uncertain environment. Isolation from family and friends, extended periods of downtime, and a lack of outlets—including school and daycare—added volatile elements to the environment. "Spending more time in the same environ-ment with others might increase the risk of conflicts between family members and could be a triggering factor for violent behavior" (Ertan et al., 2020). For victims, the crisis revealed a paradox: leaving home risked exposure to the virus but staying home risked escalating violent behavior.

Intimate partner violence

During the pandemic, DFV intensified, especially intimate partner violence—the violence by a current or former spouse or partner in an intimate relationship against the other spouse or partner (Piquero et al., 2021; Kofman and Garfin, 2020; Mlambo-Ngcuka, 2020). So much concern existed that prominent global institutions, including the WHO, UN Women, and UNICEF, issued a call to action (Piquero et al., 2021). Why did intimate partner violence intensify? As the coronavirus spread, government lockdowns limited both work and social connections. Economic, health, and social insecurities then exacerbated pre-existing conflicts. Together, these factors created a context for DFV: women and children were stuck in their homes with abusers. "It is clear that the impact of the Covid-19 pandemic will bear heavily on those navigating these unprece-dented circumstances while isolating—indefinitely—in unsafe homes" (Kofman and Garfin, 2020). Reviewing the evidence, Alex Piquero and coauthors (2021) argue:

> Combined, the stay-at-home orders as well as the economic impact of the pandemic heightened the factors that tend to be associated with domestic violence: increased male unemployment, the stress of childcare and homes-chooling, increased financial insecurity, and maladaptive coping strategies. All of these, and more, increase the risk of abuse or escalate the level of violence for women who have previous experience of violence by their male counterparts as well as violence by previously non-violent partners.

As Jane Bradley (2021) adds, government shutdown interventions "left victims trapped at home with abusers and isolated from family and friends." Sometimes the victims, such as the subject of Bradley's article, paid the ultimate price.

Unsettling outcomes

If history serves as a guide, pandemics create unsettling outcomes. Negative psy-chological effects, such as anger, confusion, and posttraumatic stress symptoms, accelerate. The closure of both schools and childcare facilities reduces house-hold flexibility, especially for mothers. Public health restrictions limit alternative sources of housing, such as shelters and safe havens. Gender equity erodes. In the household, stress results from balancing childcare, education, and work. Victims of domestic abuse, quarantined with their abusers, disconnect from their support systems and struggle to access the Internet. The breakdown of routines creates a volatile environment, fewer social connections, and less oversight. During a pandemic, many victims of DFV do not seek help (Evans et al., 2020). "Despite being a global phenomenon, (DFV) is highly underreported due to stigma and social pressures" (Mittal and Singh, 2020). Close proximity between abusers and victims limits the ability of the latter to seek assistance.

Chapter thesis and organization

The rise in DFV during the coronavirus pandemic served as an important external cost of lockdown interventions. To develop this thesis, the chapter discusses evidence, universality, diverse contexts, theories, intervention and reform, and lessons.

Evidence

During the coronavirus pandemic, countries reported an increase in DFV (Piquero et al., 2021). In Australia, coronavirus restrictions and stress caused DFV cases to spike (Kennedy, 2020). After the imposition of quarantines, China witnessed the same outcome, especially with intimate partner violence (Zhang, 2020). Calls on emergency helplines to report DFV increased by 40 percent in Brazil and 30 percent in Cyprus (Graham-Harrison et al., 2020). In India, after the implementation of a nation-wide lockdown, complaints of DFV doubled (Krishnakumar and Verma, 2021). France reported a 30 percent increase in cases (Ertan et al., 2020). In Spain, during the first 2 weeks of lockdown, an emergency hotline received 18 percent more calls relative to the previous month (Taub, 2020). In the United Kingdom, calls to the largest domestic abuse hotline soared (Townsend, 2020). In the United States, during the first year of the pandemic, cases rose (Boserup et al., 2020). Globally, the United Nations Population Fund argued that "continuing lockdowns for six months could result in an extra 31 million cases of gender-based violence" (Mahase, 2020).

Provisions

Generalizations strain the efficacy of informed observations. But, in many countries, cultural and historical practices establish that women have more family responsibilities, less policy safeguards, and fewer options for economic independence. Even in societies with relatively progressive social attitudes, DFV persists. After the first year of the pandemic, "Surveys around the world (showed) domestic abuse spiking . . . (and) jumping markedly year over year compared to the same period in 2019" (Kluger, 2021). Policy interventions initially focused on the economy/health tradeoff. An increase in unemployment accompanied an improvement in public health. But countries that fought the spread of the virus, maintained vigilance, and relied on flexible policy sustained economic recoveries. In retrospect, lockdown policies should have included provisions for DFV.

Contributing factors

Several factors link to DFV. First, in the economy, when women experience layoffs but assume greater responsibility for household chores, they become more dependent. Second, psychological problems link to DFV. A pandemic increases depressive symptoms, sleep disturbances, and substance abuse (Zhang et al., 2020). Third, due to a decline in support networks, isolation exacerbates pre-existing vulnerabilities (Boserup et al., 2020). Fourth, in the presence of a power imbalance, DFV may accelerate (Boserup et al., 2020). Fifth, in addition to phys-ical violence, methods of control include isolation from family and friends, strict rules of behavior, and surveillance (Taub, 2020). Together, the factors enhance the potential for DFV.

Universality

Rachel Snyder (2020) of American University, in *No Visible Bruises*, describes the "universality of domestic violence and how it crosses geographical, cultural, and linguistic barriers." In Snyder's framework, DFV exists as a social pandemic on a global scale. But she argues that the male of the species is far more violent and deadly than the female. In support of her claim, almost all global statistics about DFV demonstrate that men monopolize the practice: "It is men who are violent. It is men who perpetuate the majority of the world's violence, whether that violence is domestic abuse or war," and it is therefore men who have to "learn nonviolence" (Snyder, 2020). Any discussion of DFV in a nongendered manner avoids the reality of the problem. Although Snyder argues that we "could actu-ally do something about" DFV, it operates along a continuum (Snyder, 2020). Abuse may lead to family disruption on one end of the continuum, mass shoot-ings on the other, and many possibilities in between.

Interconnected economic, psychological, and social forces, such as unem-ployment, mental instability, and conflict, contribute to DFV. In the family, power dynamics may create an environment of abuse. Because these forces coalesce in the home, those who could help—counselors, social workers, teachers—often lack in-formation. The outcomes include undercounted statistics and thousands of annual homicides, "An average, in fact, of 137 women each and every day are killed by in-timate partner or familial violence across the globe. . . . And for every woman killed in the United States from domestic violence homicide, nearly nine are almost killed" (Snyder, 2020). Globally, one in three women suffer from intimate partner violence, with 42 percent of these cases leading to physical injury (Sabri et al., 2020). In the United States, according to the National Crime Victimization Survey of the Bureau of Justice Statistics, more than 3,000,000 cases and 4,000 deaths occur annually.

Home as a dangerous place

In a cycle of domestic abuse that includes controlling behavior and threats, the home serves as the most dangerous place. Victims know their abusers. But the victims may not be able to leave. A pattern of dependence may restrict their mobility. Mothers may maintain a determination to keep themselves and their children alive by any means.

> They stay in abusive marriages because they understand something that most of us do not, something from the inside out, something that seems to defy logic: as dangerous as it is in their homes, it is almost always far more dangerous to leave.
>
> *(Snyder, 2020)*

That is, leaving or discussing the act of leaving may intensify the cycle of abuse. In the presence of DFV, victims live in imperfect situations. They may not identify their own level of danger. They may not know how to situate their problem in a larger context. "They may not realize it is escalating. They may not know the specific predictors of intimate partner homicide" (Snyder, 2020). But often victims understand the severity of the situation, biding their time, protecting their children, evaluating their options, and planning their escape.

Costs of domestic violence

Domestic violence sits adjacent to other problems, including education gaps, gender and racial inequity, mental health, and poverty. But it exists as an ongoing crisis. After the identification and arrest of repeated abusers, the legal system often releases them with inadequate treatment, perpetuating the cycle. In addition to the effects on victims, the problem creates external costs in the form of broken families, lost opportunities, and lower levels of output. In the United States, "Domestic violence health and medical costs top more than $8 billion annually for taxpayers and cause victims to lose more than eight million workdays each year" (Snyder, 2020). The problem also relates to homelessness, incarceration, and homicide. Mass shootings in the United States, for example, "more than half the time, are domestic violence" (Snyder, 2020). Serving as a private and social problem, domestic violence exists as a matter of public health.

Assessment

Jacquelyn Campbell, the Ann D. Wolf Chair at the Johns Hopkins School of Nursing, created the **danger assessment** (DA), a tool that identifies the potential for domestic violence. Some of the 22 assessment factors are broad, including gun ownership and substance abuse. Others are specific, including alcohol and drug abuse, threats to children, and a victim's attempts to leave. Reflecting on Campbell's contribution, Snyder (2020) argues that the DA is the single most important tool used in intimate partner assault, treatment, and awareness today. . . . It has broken through cultural and political barriers, been adapted for use by police, attorneys, judges, advocates, and healthcare workers. It has informed research and policy and saved countless lives.

The method by which a victim answers questions determines the following steps: whether an abuser is charged with a crime or whether a victim presses charges, finds shelter, and proceeds through the legal system. The problem during a pandemic, however, is the inability to apply a DA.

Physical health effects

The physical effects of abuse are documented for women who are treated in healthcare systems (Campbell, 2002); however, it is important to emphasize that many women are not treated: "Intimate partner violence has long-term negative health consequences for survivors, even after the abuse has ended. These effects can manifest as poor health status, poor quality of life, and high use of health services." For victims, intimate partner violence is a common cause of injury, including to the face, head, and neck; however, out of fear of retaliation, many victims do not seek treatment. Intimate partner violence may create chronic health problems, including fainting, headaches, and back pain. The damage may include depression, hypertension, mental health disorders, and suppression of the immune system. Research published during the third year of the coronavirus pandemic demonstrates that victims of intimate partner violence can sustain head trauma more often than football players (Hillstrom, 2022). This research reveals the hidden pandemic of brain injury among women suffering from the problem: "But unlike injuries in sports, war or accidents, domestic assaults happen almost entirely out of view. Victims themselves may not be able to process or remember what happened, and their assaults are often not reported to the police" (Hillstrom, 2022). The implication is that, while research is addressing the scope of physical trauma, more analysis must consider how traumatic injuries affect the brain.

Immigrant women

Immigrant women constitute a special case. Immigration reform would "protect women who fear their immigration status will be used against them if they report a crime or take their partner to court" (Butcher, 2021). Bushra Sabri and her coauthors (2020), including Jacqueline Campbell, argue that many of these individuals struggle in a cycle of domestic abuse. While a lack of support, language barriers, and social stigma serve as complicating factors, a pandemic exacerbates the problem. But "support frameworks during (a) pandemic are essential to providing an adequate response to survivors of intimate partner violence, particularly those from marginalized groups" (Sabri et al., 2020). Reducing financial hardship, increasing social support, and providing mental health resources help to break the harmful cycle.

Diverse contexts

In their book, *Domestic and Family Violence*, Silke Meyer and Andrew Frost (2019) argue that society has traditionally considered conduct defined as DFV as a "private and individual matter." The prevailing thought was that it occurred in families, so society semi-tolerated the problem. But this arrangement propagated an environment of abuse. The voices of victims went unheard. The modern understanding, according to Meyer and Frost (2019), is that DFV is a "complex and diverse social issue," involving personal and social factors. While policy interventions target adults, society now recognizes the impact on children, siblings, extended family members, and friends. In the presence of external costs, DFV exists as "an issue of endemic proportions" (Meyer and Frost, 2019). For both families and society, DFV, when unresolved, damages individual lives and the fabric of communities, often for generations. For systems of criminal justice, economic activity, and public health, negative externalities persist, including crime, economic losses, and inferior health outcomes.

Economic, political, and social arrangements

Family arrangements link to cultural histories, community practices, and economic circumstances. In some households, interaction between family and social constellations escalates DFV. Despite attempts to establish solutions, the personal and social nature of the problem complicates policymaking. How may society alter the existing order, thus ending the cycle of violence? How may policy intervention counteract the need for abusers to demonstrate power and control? These nuanced and complicated questions relate to a division in theoretical perspectives.

Division in theoretical perspectives

Different perspectives offer the means to address the problem, formulate policy interventions, and establish effective solutions. One perspective, common in the literature, implies that DFV stems from a structure of inequality and oppression with roots in the systematic nature of the problem. Within society at large and households in particular, abuse arises from an imbalance of power. Patriarchal societies support hierarchy, restrict gender roles, and legitimatize the power of men. The family provides cover for both a captive environment and abusive behavior. A second perspective, also common in the literature, implies that DFV stems from a family systems arrangement. Within family dynamics, an environment of abuse may persist, existing as a private matter. But oppressive gender realities do not cause or exist as a consequence of DFV. While this perspective acknowledges domestic abuse, it does not explain why abuse persists in some households and not others, when controlling for factors such as employment and poverty.

Opposing views

The division between the structural and family systems perspectives impacts how society views the problem. While the structural perspective addresses the pattern of violence, intentional tactics of abusers, and the "agency of an abusive actor conducing a more or less deliberate regime of subjugation," the family systems perspective focuses on episodic elements of the problem in a "sequence of events contained in a bounded system" (Meyer and Frost, 2019). With the structural perspective, abusive behavior persists in a pattern of coercion and domination. That is, DFV exists as a gendered problem, informed by male domination and female oppression. "Within this framework, DFV is primarily a male-to-female perpetrated phenomenon, usually marked by the abuser's desire to strategically manipulate and control the victim" (Meyer and Frost, 2019). With the family systems perspective, abuse flows from the commission of a specific act or the omission of consent. In a family systems arrangement, DFV exists as a situational problem, compounded by financial stress, parental disagreements, and unemployment:

> Family conflict scholars argue that DFV may be used to establish or maintain status within the family structure or hierarchy but that the underlying objective is not to strategically control the victim...(but exists as) an expression of anger and frustration.
>
> *(Meyer and Frost, 2019)*

The implication of the different perspectives is that those who argue for the structural position view DFV as primarily a male-to-female problem. But those who argue for the family systems perspective view women as equally violent as men. For context, however, the data demonstrate the latter position as false. National and international statistics show that women are much more likely to suffer from DFV than men. Gender symmetry does not exist (Myhill, 2017).

The role of gender

In *Domestic Violence in Diverse Contexts: A Reexamination of Gender*, the feminist scholars Sarah Wendt and Lana Zannettino (2015) argue that "Gender affects every aspect of our lives, and violence is highly gendered." Abusers direct violence against women because they are women. Women may experience DFV from known men in their lives, particularly intimate partners:

> The risk to women from their current partners has been found to be three times greater than for men, and women are more likely to endure a wide range of violent behaviors, be injured and have a weapon used against them by their partners or ex-partners.
>
> *(Wendt and Zannettino, 2015)*

The implication is that, relative to men, women are more likely to endure multiple incidents of violence. This dynamic explains why women experience a relatively higher level of fear. Through fear, abusive men are able to control women's behavior, choices, and freedom (Yodanis, 2004).

Power and control

The problem of DFV not only exists as a means for abusers to control their victims but also serves as the most explicit form of patriarchal domination. The concept of **patriarchy**, a system in which men hold the power and women are subservient, includes modes of production, leadership, and social organization. A patriarchy establishes economic and social pressures, providing a context for DFV. Violence is a means in which an abuser establishes control. "The refusal of the state to intervene effectively in terms of welfare provision and criminal justice responses to support women is part of the problem" (Wendt and Zannettino, 2015). If vulnerable women cannot find shelter, they may continue to live with abusers. Patriarchal beliefs, methods, and social structures perpetuate the problem in the form of labor market discrimination, wage gaps, an inequality of opportunity, a lack of support in legal systems, and sexist forms of prejudice.

> In other words, men have traditionally received material/financial, emotional, social and judicial leverage due to the rights they have been afforded over time, history and culture, and hence power differences between men and women within a family can benefit the perpetrators of domestic violence who take advantage of this infrastructure.
>
> *(Wendt and Zannettino, 2015)*

The existence of a patriarchy contributes to the problem of DFV and the historical advantages of men.

Gender and patriarchal society

Domestic violence exists in the context of gendered power dynamics. Those who benefit most from the existing order may restrict the opportunities of others, especially within patriarchal contexts. While entitlement, the fact of having a right to a behavior, serves as an important factor, social attitudes and beliefs that contribute to the problem include the idea of men as the head of households and women responsible for domestic work. In this context, DFV exists within the patriarchal context in which it is constructed, relates to gendered power dynamics, and stems from "the traditions, habits and beliefs about what it means to be a man" (Wendt and Zannettino, 2015). For a society, these beliefs may become so entrenched and widespread that they become accepted.

Intersectionality

Intersectionality—the interconnected nature of social categorizations such as race, class, and gender—describes how layers of disadvantage concentrate the effects of abuse. In the United States, about 25 percent of women have been victims of physical violence, sexual violence, or stalking, the majority first experiencing these problems before the age of 25 (Butcher, 2021). While these problems exist for white women, the risk of violence is greater for women of color. During their lifetime, in the United States, more than 40 percent of Black women experience domestic violence; 56 percent of Native American or Native Alaskan women experience intimate partner violence (Butcher, 2021).

> In the context of domestic violence, intersectional theory has emphasized the need to move away from a generalization or homogenization of women's experiences of violence and oppression, to an understanding of how the diversity amongst women as a group will engender different experiences, contexts and responses to violence.
>
> *(Wendt and Zannettino, 2015)*

When attempting to address the problem, women with multiple forms of marginalized status struggle to find effective solutions.

Barriers

Intersectionality elucidates the ways in which the intersections between gender, race, and socioeconomic status present barriers. With DFV, an intersectional focus reveals a range of correlating factors. It also describes the help-seeking arrangements of victims and how these arrangements may differ across different settings. In the dominant social class, victims establish certain arrangements. But for victims with marginalized status, help-seeking arrangements may not exist. According to an article in *The New England Journal of Medicine*, the pandemic "reinforced important truths: inequities related to social determinants of health are magnified during a crisis and sheltering in place does not inflict equivalent hardship on all people" (Evans et al., 2020). While gender provides the most important context for DFV, it exists along a range of intersecting factors, including race, class, and socioeconomic status. During a pandemic, individuals with multiple forms of marginalized status suffer. Economic instability, a lack of childcare, neighborhood violence, and unsafe housing worsen the problem.

Layers of disadvantage

By addressing layers of disadvantage, intersectionality elucidates the incidence of DFV:

> Much of the success of the movement against domestic violence has been assisted by the feminist framing of domestic violence which reiterates an unambiguous and straightforward message that domestic violence is common, dangerous to women, and affects women of all social standing, effectively cutting across stratifications of ethnicity and socioeconomic status.
>
> *(Nixon and Humphreys, 2010)*

While DFV exists as a problem across ethnic stratifications and socioeconomic status, intersectionality reinforces the idea that gender serves as an important area of focus.

A framework

The point is that intersectionality establishes a framework to analyze individual experiences, acknowledging the diversity of women's lives. This approach reveals how characteristics of a family's internal dynamics and the social environment add layers of complexity to the problem. For example, a mother's dependence on an abusive partner during a pandemic exacerbates the difficulties of living in poverty. Because of the presence of poverty conditions, the victim experiences a higher level of entrapment. Even more, social categorizations such as ethnicity and race may combine with poverty and dependence to limit access to resources, including counseling, income assistance, and social services.

Case study 6.1 Forced migration, gender violence, and Covid-19

Large-scale conflicts in Syria, Honduras, Yemen, and other countries, characterized by civil war, interstate conflict, and insurrection, lead to humanitarian crises. The effects are widespread. Individuals are displaced from their homes. Economies crumble. The food supply dwindles. Deteriorating social conditions force mass migration. But where do forced migrants go? The answer is anywhere they will be accepted.

One problem is the inability to accommodate forced migrants. Political and social realities dictate a limited set of options. An inability to travel long distances may leave forced migrants in transit. They may reach migrant camps in adjacent countries, where they have to wait for permanent settlement. Until recently, Canada, the United Kingdom, the United States, and members of the European Union (EU) served as common destinations, but these liberal democracies have become more reluctant to accommodate forced migrants. For example, the EU contracts with Turkey to handle forced migrants from Africa and the Middle East, establishing a place of refuge, transit, and resettlement.

Another problem involves living conditions, such as crowding and poor access to food, healthcare, and sanitary environments. In migrant camps, sexual and gender-based violence (SGBV) against women and children festers. The high levels of SGBV in both transit and migrant camps highlights the ongoing tragedy: "SGBV includes rape and sexual assault, as well as physical, psychological or emotional violence; forced marriage; forced sex work; and denial of resources, opportunities, services and freedom of movement on the basis of socially ascribed gender roles and norms" (Phillimore et al., 2021). Forced migrants move through difficult topography, leading to forms of control that are not tied to place.

An intersectional framework highlights the reality that violence against women and children stems from multiple systems of control, inequality, and oppression. Migrant women and children face ongoing disadvantage from their gender, legal status, race, religion, and other forms of marginalization. This framework emphasizes subordination and vulnerability, stemming from the precarious nature of forced migration. Problems include immigrant status, prejudice, and stereotypes. In the presence of a harmful, threatening, and uncertain set of circumstances, individuals struggle to seek asylum.

Migrant camps often entail inadequate medical care, overcrowding, and squalid conditions. The threat of SGBV exacerbates the situation. "Forced migrant women and their children report family violence throughout the process of resettlement and struggle to access services that do not account for the complexities of forced migrants' SGBV experiences" (Phillimore et al., 2021). Because the full incidence of SGBV is unreported, the true extent of the problem is unknown; however, it is clear that the coronavirus pandemic exacerbated the problem.

In the case of forced migrants, humanitarian advocates argue for comprehensive programs of assistance. "However, the scale of recent emergencies has not been matched with the appropriate resources, capacity, political will, or governance models to enable the development of gender-sensitive services and facilities" (Phillimore et al., 2021). Countries often abandon forced migrants and lack the capacity for accommodation. The presence of Covid-19 contributed further complexity to the intersecting factors of inequality, oppression, and abuse.

Enacting violence in private spaces

Responses to DFV must account for both the complex nature of the problem and the centrality of family life. Because DFV exists in private space, public policy must consider actions, intentions, and responses. However, to address the persistence of the problem, it is important to consider the nature of offenders and why they cling to power, suffer from insecurity, and inflict physical and psychological harm on others. This is a challenging task. The oppressor often shares experiences with the victim, maintaining a (flawed) relationship. In addition, abuse exists in a household environment. The oppressor may express contrition, distress, and remorse—truthfully or not—complicating the process of intervention. To assist victims and establish solutions, it is important to consider several factors, including the systematic nature of insecurity, power differentials, actions and motivations, and the culture of **masculinity** (the qualities or attributes regarded as characteristic of men).

Systematic nature of insecurity

The systematic nature of insecurity plays a role. A lack of education, initiative, and skills may restrict employment opportunities. Attempting to address the problem, inadequate institutional support may prevent both recognition and action. An inability of family, friends, or colleagues to intervene may help to perpetuate the problem.

Power differentials

Violence or the threat of violence normally results from a desire to control. It is this behavior that helps to delineate an environment of abuse. Studies in feminist analysis, grounded in the concept of power differentials between women and men, demonstrate that gender inequality serves as the basis for both patriarchal societies and the persistence of DFV:

> To achieve a fundamental grasp of the perpetrator's sense of entitlement in using a range of abuse tactics against those he lives with (and professes to care for), we must be deeply aware of the breadth of convention and the depth of institutional bedrock that support this behavior. This involves understanding his access to privileges and power and the features of society that effectively turn a blind eye to, or are even collusive with, such conflict.
>
> *(Meyer and Frost, 2019)*

This approach, when effective, acknowledges the existence of a range of actions and motivations.

Actions and motivations

A range of actions and motivations contributes to violent behavior. This range persists across cultural groups, family arrangements, and social settings, making clear why DFV serves as a global and complex problem. The monolithic image of the serially abusive and exploitive oppressor does not capture nuance. While a stereotype contributes to arrests and convictions, oppressive behavior often results from anger, frustration, and impulse. The existence and realization of both systematic and situational violence informs both the assessment of the problem and appropriate tools of policy intervention (Meyer and Frost, 2019).

Culture of masculinity

In a shared system of meaning, a culture of masculinity demonstrates the qualities or attributes of men. Through behavior and conduct aligned with customs, expectations, and traditions, individuals demonstrate masculinity:

> Under these circumstances, women are more readily objectified, subjugated, and exploited by men who themselves are attempting, sometimes desperately, to conform to the demands of highly prescriptive blueprints for masculinity. Especially in contexts of disadvantage and trauma, arguably such demands often outstrip coping resources and responses that would allow for non-violent outcomes.
>
> *(Meyer and Frost, 2019)*

Masculine hierarchies reveal the nature of bravery, physical prowess, and risk-taking. These factors are reflected in employment, opportunity, and status. When men compete along these characteristics of masculinity, socially-sanctioned conflict may result. When this motivation pervades cultural expectations, individuals may convey the idea of control to maintain a hierarchical position. During a crisis, an abuser may resort to aggressive or violent tactics, an attempt to keep family members in subservient positions (Meyer and Frost, 2019).

Resisting violence in private space

The creation of a victim occurs over time. Normally the process entails sustained and prolonged levels of abuse. Although specific circumstances differ, DFV may exist in the presence of economic stress, power and control, and mental illness. A crisis such as a pandemic exacerbates the problem. In this context, the action of a victim exists as learned behavior, living with an abuser's outbursts because the alternative is viewed as inferior. The victim may try different tactics, either maintaining vigilance with the prospect of ending the abuse or realizing that the tactics do not work. In this dynamic process, marginal success spurs further attempts to mitigate the problem. But isolation, inadequate resources, and insufficient support may discourage further attempts to end the violence. A victim, in other words, may exhibit agency, establish contingencies, and struggle to identify a future pathway, behaviors that may remain unknown to society (Meyer and Frost, 2019).

Construction of DFV victims

The characteristics of DFV reveal a gendered problem, a difference in physicality between abuser and victim, and the propagation of family roles. Social norms may dictate that the maintenance of family structure trumps family dissolution. In this context, the victim may exhibit the role of mother, wife, and partner. However, society may possess the misguided view that victims share in the responsibility for DFV, even though the responsibility exists with the abuser. The misguided view stems from the misperception of choice: victims should either avoid the relationship or leave after the abuse begins. But abuse may begin during a relationship, escalate in a crisis, and prevent the establishment of better circumstances. Victims normally find themselves experiencing this toxic environment through no fault of their own (Meyer and Frost, 2019).

Stigma

Stigma, a mark of disgrace, correlates with cultural norms, social views, and personality traits. Examples include addiction and criminal behavior. In the presence of stigma, society may engage in the practice of victim-blaming. When seeking support, victims may not receive an appropriate level of attention. In the presence of lockdown orders, society may not address the problem. When establishing social services, an insufficient level of resources for hotlines and shelters may exist. Stigma may partially explain why insufficient responses during the pandemic propagated DFV. (The severity of the health crisis, a lack of information, economic contraction, family uncertainty, and insufficient preparedness serve as other factors.) While victims are not helpless, appropriate responses require an understanding of the intricacies of the problem, the reality of personal differences, and acknowledgment of the possibility of stigma. Victims do not have to redeem themselves. Abusers must stop the cycle of violence (Meyer and Frost, 2019).

Inability to leave

A victim may not be able to leave an abuser. Rational choices may influence the decision. That is, victims may evaluate the available information, consider alternatives, and make informed decisions. They consider risks and rewards. But the cost-benefit calculus involves nuance. Domestic violence and its contributing factors, such as manipulation, threats, and strategic control, exist alongside the safety of children, economic livelihood, and future prospects. Because of the systematic nature of the problem, the controlling behavior of the abuser, and a lack of alternatives, the victim may stay: she may perceive the costs of leaving as exceeding the benefits. Even more, the act of leaving or planning to leave may serve as the period of time associated with the greatest level of risk. Abusers may understand the gravity of the situation. They may lash out. Because of previous experience, victims may understand this possibility. In the presence of instability, it is therefore important to consider this volatile reality (Meyer and Frost, 2019).

Theories

Studies on the determinants of domestic violence reveal multiple factors. But variation exists with respect to the prevalence and severity of the problem. According to Christine Arthur and Roger Clark (2009), in an influential article, societal-focused theories demonstrate different viewpoints, including the following theories.

Culture of violence

The **culture of violence theory** posits that violent societies are more likely than nonviolent societies to experience conflict. In this context, violent societies are also more likely to permit domestic violence, because it is accepted as a means of household conflict resolution. The idea is that social acceptance of the problem blocks methods of prevention, creating an environment of abuse. The implication is that, as society engages in conflict, domestic violence exists as a repeated action within the household (Arthur and Clark, 2009).

Dependency

Economic dependency theory posits that this dependency lessens a partner's status. Economic entanglement and a lack of alternative income sources may complicate the process of divorce. In addition, economic dependency correlates with less access to economic, educational, and social resources. Capitalism, which entails a market-based system of exchange, penetrates the social fabric of the family, creating an environment of dependency. Because of the dependency of the victim, family dynamics may lead to abuse. In countries that rely on capitalist structures, corporate interests may overwhelm social reforms, limiting the effectiveness of policies that fight DFV (Arthur and Clark, 2009).

Exchange

Exchange theory posits that domestic violence persists in the presence of perceived net benefits for abusers (benefits minus costs). Even if misplaced, the perceived benefits to abusers include control, influence, and power. The costs to abusers include family disfunction, legal consequences, and personal strife. In the minds of abusers, when will net benefits exist? In the presence of sexism and entrenched patriarchy, the perceived benefits to the abuser are more likely. In this context, low or nonexistent costs of DFV stem from weak or absent cultural norms, public policies, and social pressures that prevent DFV. The theory implies that progressive reforms may increase the costs and decrease the benefits of DFV (Arthur and Clark, 2009).

Modernization

Modernization theory posits that enhancing women's status through labor market participation reduces income gaps between men and women and undermines the patriarchy. But challenges remain. The influx of women into the labor force may not reduce income gaps. Because mothers are normally responsible for household duties, including childcare, progressive social reforms must accompany economic reforms, such as early-childhood education and daycare. A modern society frees women (and men) from traditional gender roles. As traditional gender roles decline, DFV decreases (Arthur and Clark, 2009).

Patriarchy

Patriarchal theory posits that gender imbalances propagate lower social status for women. In this context, patriarchal norms perpetuate an environment of abuse. With historical roots, this cultural belief emphasizes subordination and control. But modern cultures may dismantle the patriarchy. Economic policies may decrease income gaps between men and women. The judicial and legal systems may enhance women's freedom. Social norms may encourage the participation of women in positions of power and influence. Over time, the accumulation of these effects weakens the patriarchy (Arthur and Clark, 2009).

Resources

Resource theory posits that the resources of the main income earner correlate with power. But more resources for the main income earner lead to less domestic violence. If power serves as an organizing principle, resources for dependents poses a threat to the existing order. In this context, DFV serves as a method to maintain a traditional power dynamic. When power erodes, methods to reassert power increase, including DFV (Arthur and Clark, 2009).

Intervention and policy reform

Domestic and family violence reflects a collective failure to both treat the problem and assign it pandemic status. During Covid-19, the acceleration of DFV highlighted this reality. When society established shutdown conditions, the problem escalated. But to eliminate this shadow pandemic, society must dismantle the factors that perpetuate DFV. "The pandemic has highlighted how much work needs to be done to ensure that people who experience abuse can continue to obtain access to support, refuge, and medical care when another public health disaster hits" (Evans et al., 2020).

Strategies

At the level of government, policymakers must address the social determinants of DFV, establishing appropriate standards of care. In patriarchal societies, policy reforms are difficult to implement, highlighting the need for more women in positions of leadership. The impact of DFV on victims depends on access to resources, community help, and the ability of victims to extricate themselves. Preventive measures must include the expansion of hotline services, while acknowledging the difficulty of access during crises. The justice system must provide a functional legal environment, especially during periods of lockdown. On social media, public awareness campaigns must provide education on the ubiquity of the problem, urging authorities to act.

Additional resources

At the community level, social services require resources, especially well-trained staff, to identify the problem and code the cases without alerting abusers. At the level of the household, Wi-Fi technology enables wider access to support. Because victims may hesitate to report abuse, a stable and reliable system of communication must exist. With a sufficient number of resources, medical professionals must identify victims, provide counseling, and connect with social services. In these settings, the evaluation of DFV cases provides the capacity for intervention, including a review of safety planning. Amy Butcher (2021) argues:

> We also can improve how we respond to women in distress. Our law enforcement officers need further training so that they can better recognize body language and behavior indicative of ongoing emotional, verbal and physical abuse. A clinical social worker should also be a part of the team that responds to reports of domestic violence, to help de-escalate conflicts and guide those in danger to safe shelters.

Society must bolster first responder networks with more staff, operational capacity, and training. For victim safety, communities must provide access to shelters and housing (Evans et al., 2020).

Safety of victims

Society must guarantee the safety of victims. If social norms suggest a "sanctity of family life," the private nature of DFV complicates the ability of victims to both speak out and leave abusive situations. It is, therefore, crucial to critically evaluate idealized representations of family. In the present era, millions of women and children suffer from DFV, and, often, suffer secretly. In the presence of abuse, victims must be able to evaluate their situations and leave volatile environments. If victims are able to extricate themselves, they must find assistance, secure housing, and establish childcare. During stable periods, these are challenging tasks. In the presence of a crisis, they become prohibitive.

Risk factors

Together, these methods of intervention and policy reform demonstrate the need to strengthen legal methods of oversight, support systems, and society's ability to combat DFV. Protecting victims requires the reduction of risk factors for violent behavior, including economic instability, social insecurity, and a culture of dependence. During the pandemic, many governments, healthcare practitioners, and nonprofit organizations took steps to address these problems, increasing staffing for hotlines, maintaining family contacts, and publicizing the problem of DFV. But during times of stability, a substantial amount of work remains (Ertan et al., 2020; UNFPA 2019).

Vulnerability

Cultural norms, economic arrangements, and social expectations may perpetuate high levels of vulnerable for some women and children. The reason is that

> home can be a place where dynamics of power can be distorted and subverted by those who abuse, often without scrutiny from anyone "outside" the couple, or the family unit. In the Covid-19 crisis, the exhortation to stay at home therefore has major implications for those adults and children already living with someone who is abusive or controlling.
>
> *(Bradley-Jones and Isham, 2020)*

But with current research, knowledge, and documentation of DFV, it is unacceptable for society to tolerate the problem anymore.

Lessons

To protect public health during times of crisis, society relies on frontline workers, including health professionals, nurses, and social workers. For these essential employees, society must provide both protection and resources. But society should also support complementary workers, the "advocates, therapists and helpline practitioners working in specialist domestic and sexual violence services in the voluntary sector" (Bradley-Jones and Isham, 2020). During crises, complementary services provide advocacy, mentoring, peer support, and refuge. The independence of complementary services exists as an important feature, especially in the presence of extenuating circumstances, such as difficult interactions with police. For workers, examples include childcare, living wages, and a stable supply of personal protective equipment. For victims, examples include opportunities to contact help without alerting abusers, more capacity for hotline services, and information campaigns (Bradley-Jones and Isham, 2020).

Policy design

In preparing for future crises, policymakers must weigh the health benefits of policy intervention against the economic and social costs of lockdown interventions. That is, policymakers must address the overall implications of policy design, including unintended outcomes. Policy that intends to improve public health should not intensify DFV. In an intersectional framework, policymakers should pay specific attention to individuals with multiple forms of oppression, such as those suffering from both poverty and discrimination. Policymakers should also focus on individuals who are isolated, including the elderly, people with mental health or chronic conditions, and women and children suffering from abuse. It is important to emphasize that, for victims, the most dangerous and lethal time is the period in which they are trying to leave. In these situations, they are at the highest risk for bodily injury or homicide.

Flow of information

The rise in DFV during the coronavirus pandemic underscores the need to increase the flow of information on the problem. As the pandemic surged, social and health care workers warned of a rise in DFV, given the increased levels of isolation. The interconnected factors of the spread of the virus, economic collapse, increase in morbidity and mortality, lack of face-to-face schooling, mental health toll, and rising unemployment exacerbated the problem.

Homicides

One of the most difficult manifestations of DFV is homicide. In the United Kingdom, for example, about two women are killed every week by current or ex-partners (Bradley-Jones and Isham, 2020). In Spain, during the pandemic, the first report of a fatality from DFV came 5 days after initial lockdown, but the woman was the country's seventeenth murder victim from DFV in 2020 (Higgins, 2020). Even though homicides are a small percentage of DFV cases, they demonstrate both the seriousness of the problem and the urgency in ending the cycle of violence.

Pre-existing problems

Crises exacerbate pre-existing problems. During a crisis, if policymakers address health consequences but ignore other outcomes, the problem of DFV worsens.

Policymakers must, therefore, anticipate this possibility, make bold statements about prevention, and create guidelines that expand victims' access to health and support services. Society must anticipate the need for social services, acknowledging that DFV occurs in private spaces. As Piquero et al. (2021) argue:

> The increase may include reports by a new set of domestic violence victims whose violence experiences are largely a function of the current economic impact of the pandemic, as well as the temporary isolation resulting from social distancing measures and stay-at-home orders. . . . The pandemic may have also served as the catalyst for those who were victims prior to the pandemic to report their experiences due to the increased incidence and severity of violence by their previously abusive partners.

Therefore, while the evidence reveals the problem of DFV, future research must include clinical data, reports from shelters and social support agencies, and information from hotline centers before, during, and after a crisis. As Kofman and Garfin (2020) conclude, "Long-term, the pandemic may serve as a critical inflection point for implementing planning and preparedness guidelines to protect (DFV) victims . . . (in) future disasters."

Restrictions on movement

For victims, restrictions on movement reduce or eliminate pathways of escape, opportunities for support, and methods of coping. In the presence of coercion, control, and surveillance, restrictions contribute to abusive environments. The reason is the context of private space within family and intimate relationships. As an unintended outcome, lockdown measures grant abusers greater freedom and opportunity to act with impunity (Bradley-Jones and Isham, 2020).

Safeguards

It is imperative to strengthen safeguards. Disasters will strike. Economic insecurity will plague low-income households. Social services will struggle to allocate scarce resources. The public sector must strengthen frontline institutions, including crisis centers, health departments, and shelters. Mental health agencies must mitigate the psychological damage of victims. After a crisis, society must address a surge in posttraumatic stress disorder, a common mental health aftereffect.

Momentum

As a shock that assumed the status of global crisis, the coronavirus pandemic contributed to a rise in DFV. But a power imbalance in families perpetuated the problem, stemming from discrimination, gender stereotypes, inequality between women and men, social norms, and societal structures. If society intends to eradicate DFV, it must establish a comprehensive policy framework to address these causes, deconstructing the reasons that society tolerates the problem. This is an important task. Even though substantial evidence reveals the seriousness of the problem, dismantling the attitudes, beliefs, stereotypes, norms, and structures that perpetuate DFV requires a long-term commitment. Society must build on previous momentum, because the global cost of DFV is countless ruined lives (UN Women, 2020). Vaeza's (2020) assessment includes several recommendations:

- Eliminate the impunity of abusers.
- Ensure the viability of support services.
- Increase funding for women's rights groups and defenders of human rights.
- Invest in prevention.
- Mobilize communities.
- Promote behavioral changes.
- Promote healthy and equitable relationships.
- Recognize the effects of DFV on women and children.
- Work with men and boys to transform harmful masculinities and patriarchies.

Considering the evidence, there is no reason to believe that DFV is inevitable. Society must eliminate the shadow pandemic of DFV.

Summary

During the coronavirus pandemic, a shadow pandemic of domestic and family violence wreaked havoc, operating as an opportunistic infection. During the lockdown interval, victims could not go outside, seek help from extended family or friends, or visit social services. With an increase in demand for support, institutions that were charged to protect women and children from DFV, already underfunded, buckled with a shortage of resources. As the pandemic progressed, many governments addressed the problem, but they could not prevent a rise in DFV. In the presence of crisis, domestic violence correlates with confinement. Government interventions that reduce disease contagion, including quarantines and sheltering-in-place requirements, exacerbate the problem. But DFV crosses cultural, geographical, and linguistic barriers, serving as a universal problem in all socioeconomic contexts. While substantial evidence demonstrates the nature and prevalence of DFV, understanding the factors that contribute to violence in private spaces informs the process of policy intervention. Societal-focused theories provide a framework of analysis, including culture of violence theory, economic dependency theory, exchange theory, modernization theory, patriarchal theory, and resource theory. Intervention and reform must consider the collective failure to both treat DFV and assign it the status of an ongoing pandemic. The lessons of DFV, including complementary services, policy design, the flow of information, homicides, pre-existing problems, restrictions on movement, and safeguards, point to the momentum created during the coronavirus pandemic to eliminate the problem.

Chapter takeaways

LO1   During the coronavirus pandemic, the shadow pandemic of DFV escalated.

LO2   Government measures that curbed the spread of the coronavirus also increased DFV.

LO3   DFV exists as a universal problem, crossing cultural, geographical, and linguistic barriers.

LO4   DFV involves individuals who are bound together by a range of personal and social factors.

LO5   Societal-focused theories provide a framework to analyze DFV.

LO6   The elimination of DFV requires bold action plans, policies, and preventive measures.

LO7   Lessons of the shadow pandemic include complementary services, policy design, the flow of information, homicides, pre-existing problems, restrictions on movement, safeguards, and momentum.

Key terms

Culture of violence theory
Danger assessment
Economic dependency theory
Exchange theory
Masculinity

Modernization theory
Patriarchy
Patriarchal theory
Resource theory
Shadow pandemic

Questions

1   During the coronavirus pandemic, explain why the shadow pandemic of DFV escalated.

2   What is the evidence of a rise in DFV during the coronavirus pandemic?

3   In what sense is DFV a universal problem?

4   How does the concept of intersectionality inform the discussion of DFV?

5   Why does the implementation of public policies to fight DFV require a better understanding of both the enactment of and resistance to DFV?

6   List and explain the theories of domestic violence. With respect to a large-scale crisis such as a disease pandemic, which theories are most relevant? Explain your answer.

7   What are the factors that impact the processes of intervention and policy reform?

8   With respect to DFV during the coronavirus pandemic, what lessons are most important? Does momentum in solving the problem exist? Why or why not?

References

Anastario, M., Shehab, N., and Lawry, L. 2009. "Increased gender-based violence among women internally displaced in Mississippi 2 years post-Hurricane Katrina." *Disaster Medicine and Public Health Preparedness*, 3(1): 18–26.

Arthur, Christine and Clark, Roger. 2009. "Determinants of domestic violence: a cross-national study." *International Journal of Sociology of the Family*, 35(2): 147–167.

Boserup, Brad, McKenney, Mark, and Elkbuli, Adel. 2020. "Alarming trends in US domestic violence during the Covid-19 pandemic." *The American Journal of Emergency Medicine*, 38: 2753–2755.

Bradley, Jane. 2021. "A Yearlong Cry for Help, then Death after an Assault." *The New York Times*, August 10.

Bradley-Jones, Caroline and Isham, Louise. 2020. "The pandemic paradox: the consequences of Covid-19 on domestic violence." *Journal of Clinical Nursing*, 29: 2047–2049.

Butcher, Amy. 2021. "I Know all too Well How a Lovely Relationship Can Descend into Abuse." *The New York Times*, September 30.

Campbell, Jacquelyn. 2002. "Health consequences of intimate partner violence." *The Lancet*, 359: 1331–1336.

Ellsberg, M. and Heise, L. 2005. *Researching Violence Against Women: A Practical Guide for Researchers and Activists*. Washington, DC: World Health Organization.

Ertan, Deniz, El-Hage, Wissam, Thierreec, Sarah, Javelot, Herve, and Hingray, Coraline. 2020. "Covid-19: urgency for distancing from domestic violence." *European Journal of Psychotraumatology*, 11(1): 1–6.

Evans, Megan, Lindauer, Margo, and Farrell, Maureen. 2020. "A pandemic within a pandemic—intimate partner violence during Covid-19." *The New England Journal of Medicine*, 383: 2302–2304.

Graham-Harrison, E., Giuffrida, A., Smith, H., and Ford, L. 2020. "Lockdowns Around the World Ring Rise in Domestic Violence." *The Guardian*, March 28.

Higgins, Natalie. 2020. "Coronavirus: When Home Gets Violent under Lockdown in Europe." *BBC News*, April 13.

Hillstrom, Christa. 2022. "Why Can't I See Straight?" *The New York Times Magazine*, March 6.

Kennedy, Else. 2020. "'The Worst Year': Domestic Violence Soars in Australia during Covid-19." *The Guardian*, November 30.

Kluger, Jeffrey. 2021. "Domestic Violence is a Pandemic within the Covid-19 Pandemic." *Time*, February 3.

Kofman, Yasmin and Garfin, Dana. 2020. "Home is not always a haven: the domestic violence crisis amid the Covid-19 pandemic." *Psychological Trauma*, 12(supplement): S199–S201.

Krishnakumar, Akshaya and Verman, Shankey. 2021. "Understanding domestic violence in India during Covid-19: a routine activity approach." *Asian Journal of Criminology*, 16(1): 1–17.

MacDonald, Rhona. 2005. "How women were affected by the Tsunami: a perspective from Oxfam." *PLoS Medicine*, 2(6): 474–475.

Mahase, Elisabeth. 2020. "Covid-19: EU states report 60% rise in emergency calls about domestic violence." *BMJ*, 369: 1.

Meyer, Silke and Frost, Andrew. 2019. *Domestic and Family Violence*. New York: Routledge.

Mittal, Shalini and Singh, Tushar. 2020. "Gender-based violence during Covid-19 pandemic: a mini review." *Frontiers in Global Women's Health*, 1(4): 1–7.

Mlambo-Ngcuka, Phumzile. 2020. "Violence Against Women and Girls: the Shadow Pandemic." *United Nations*, April 6.

Myhill, Andy. 2017. "Measuring domestic violence: context is everything." *Journal of Gender-Based Violence*, 1(1): 33–44.

Nixon, Jennifer and Humphreys, Cathy. 2010. "Marshalling the evidence: using intersectionality in the domestic violence frame." *Social Politics*, 17(2): 137–158.

Parkinson, Debra and Zara, Claire. 2013. "The hidden disaster: domestic violence in the aftermath of natural disaster." *Australian Journal of Emergency Management*, 28(2): 28–35.

Phillimore, Jenny, Pertek, Sandra, Akyuz, Selin, Darkal, Hoayda, Hourani, Jeanine, McKnight, Ozcurumez, Saime, and Taals, Sarah. 2021. "'We are forgotten': forced migration, sexual and gender-based violence, and Coronavirus disease-2019." *Violence Against Women*, 28(9): 2204–2230.

Piquero, Alex, Jennings, Wesley, Jemison, Erin, Kaukinen, Catherine, and Knaul, Felicia. 2021. "Domestic violence during the Covid-19 pandemic – evidence from a systematic review and meta-analysis." *Journal of Criminal Justice*, 74: 101806.

Rubenstein, B., Lu, L., MacFarlane, M., and Stark, L. 2020. "Predictors of interpersonal violence in the household in humanitarian settings: a systematic review." *Trauma, Violence & Abuse*, 21(1): 31–44.

Sabri, Bushra, Hartley, Maria, Sahab, Jyoti, Murray, Sarah, Glass, Nancy, and Campbell, Jacquelyn. 2020. "Effects of Covid-19 pandemic on women's health and safety: a study of immigrant survivors of intimate partner violence." *Health Care Women International*, 41(11–12): 1294–1312.

Snyder, Rachel. 2020. *No Visible Bruises*. New York: Bloomsbury.

Taub, Amanda. 2020. "A New Covid-19 Crisis: Domestic Abuse Rises Worldwide." *The New York Times*, April 6.

Townsend, Mark. 2020. "Shocking New Figures Fuel Fears of more Lockdown Domestic Abuse Killings in UK." *The Guardian*, November 15.

United Nations Fund for Population Activities (UNFPA). 2019. *The Inter-Agency Minimum Standards for Gender Based Violence in Emergencies Programming*. New York: United Nations.

UN Women. 2020. *Covid-19 and Ending Violence Against Women and Girls*. New York: United Nations.

Vaeza, M. 2020. Addressing the Impact of the Covid-19 Pandemic on violence Against Women and Girls. *UN Chronicle*, November 27.

Wendt, Sarah and Zannettino, Lana. 2015. *Domestic Violence in Diverse Contexts: A Reexamination of Gender*. New York: Routledge.

World Health Organization (WHO). 2021. *Violence Against Women Prevalence Estimates*. Geneva: World Health Organization.

Yodanis, C. 2004. "Gender inequality, violence against women, and fear: a cross-national test of the feminist theory of violence against women." *Journal of Interpersonal Violence*, 19(6): 655–675.

Zhang, Hongwei. 2020. "The influence of the ongoing Covid-19 pandemic on family violence in China." *Journal of Family Violence*, 4: 1–11.

Zhang, Jie, Lu, Huipeng, Zeng, Haiping, Zhang, Shining, Du, Qifeng, Jiang, Tingyun, and Du, Baoguo. 2020. "The differential psychological distress of populations affected by the Covid-19 pandemic." *Brain, Behavior, and Immunity*, 87: 49–50.

7
EPISTEMIC INJUSTICE

Chapter learning objectives

After reading this chapter, you will be able to:

LO1 Explain the Capitol insurrection in the context of epistemic injustice.
LO2 Define epistemology as the study of the nature of knowledge.
LO3 Address why an infodemic hampers a reliable and effective response to a crisis.
LO4 Evaluate types of epistemic injustice.
LO5 Link intersectionality with epistemic injustice.
LO6 Demonstrate that epistemic injustice delineates a distinctive class of wrongs.
LO7 Analyze methods to address the problem of epistemic injustice.
LO8 Discuss chapter lessons.

Chapter outline

Insurrection
Epistemology
Infodemic
Epistemic injustice
Intersectionality
Evolving concepts
Addressing epistemic injustice
Lessons
Summary

DOI: 10.4324/9781003310075-9

Insurrection

Storming the Capitol

On January 6, 2021, in Washington, D.C., speaking on a podium to a large and restless crowd outside the White House, the president of the United States, Donald J. Trump, after losing his bid for reelection the previous November, declared to his supporters: "We will never give up. We will never concede. It doesn't happen. You don't concede when there's theft involved. . . . And after this, we're going to walk down, and I'll be there with you . . . we're going to walk down to the Capitol. . . . " (Blake, 2021). Trump's propagation of **disinformation**, a deliberate act of deception—the false belief that the election was stolen from him—led to a historic attack:

> Glass shattered, and a dark-clothed man climbed over the shards of a broken window and leapt down like a cat burglar to the polished floor. The moment, at about 2:13 in the afternoon, marked the first sustained breach of the Capitol since a fiery attack by the British in 1814—only this time, the attackers were American. Other insurrectionists followed, including one wielding a bat and another holding a Confederate flag. A locked door was kicked open, other windows were smashed and the rioters rushed in.
>
> *(Barry et al., 2021)*

Over the next 2 hours, thousands of Trump's followers, in the hallowed chambers and halls of the Capitol, attempted to overthrow the election, while pictures and streaming videos captured the "hostility and fear, the valor and violence—the shocking but ultimately failed attempt to derail the republic's democratic process in the name of Donald J. Trump" (Barry et al., 2021). Trump's rioters scaled walls, climbed in windows, broke down doors, invaded offices, and clashed with Capitol police. Some of the police officers succumbed to mob attack, while others prevented rioters from wreaking further havoc: "Sprayed chemicals choked the air, projectiles flew overhead and the unbridled roars formed a battle-cry din—all as a woman lay dying beneath the jostling scrum of the Jan. 6 riot" (Barry et al., 2021). While the insurrectionists had ties to rightwing extremist groups, attempting to prevent the transfer of power, many of the rioters were ordinary members of Trump's faithful: "They largely represented a group certain to have powerful sway in the nation's tortured politics to come: whiter, slightly older and less likely than the general voting population to live in a city or be college-educated" (Barry et al., 2021). But many clung to the false belief that the election was stolen. On that fateful day, while some intended to inflict harm, others did not plan to trespass or riot, but they were caught in the moment. After several hours, the tally was grim: hundreds experienced physical injuries but five individuals from different backgrounds and parts of the country who listened to Trump's disinformation lost their lives.

Electoral realities

The law of the United States defines insurrection as "a violent uprising by a group or movement acting for the specific purpose of overthrowing the constituted government and seizing its powers." On January 6, 2021, the insurrectionists among Trump's followers attempted to overthrow a democratically elected incoming government. But they had three problems. First, on the ground, there was no way for the insurrectionists to alter the election. Second, in the legal system, challenges of the election results by Trump's team did not hold up in court. Third, the election occurred in a fair and responsible manner. That is, the election was secure (Brennan Center for Justice, 2020).

Truth prevails

Because of the fair, responsible, and secure nature of the 2020 election, the electoral college totals held: 306 votes for Joseph P. Biden, former Vice President of the United States, and 232 votes for Trump. Biden became president. But damage from the insurrection was done. After the Capitol riots, many people lost faith in the sanctity of U.S. elections. Others worried about a lasting domestic insurgency. The images of rioters occupying the Capitol building tainted Trump's already mixed legacy, leading to a second impeachment. But many of Trump's followers continued to argue for a stolen election, despite evidence to the contrary (Pennycook and Rand, 2021). Subsequent reporting on the riot, however, revealed something even worse:

> The country was hours away from a full-blown constitutional crisis—not primarily because of the violence and mayhem inflicted by hundreds of President Donald Trump's supporters but because of the actions of Mr. Trump himself. In the days before the mob descended on the Capitol, a corollary attack—this one bloodless and legalistic—was playing out down the street in the White House, where Mr. Trump, Vice President Mike Pence and a lawyer named John Eastman huddled in the Oval Office, scheming to subvert the will of the American people by using legal sleight-of-hand.
> *(Editorial, 2021)*

Eastman, who was serving as a legal advisor to Trump, was propagating the idea that, on the day of electoral vote certification, January 6, 2021, the vice president should reject dozens of Biden's certified electoral votes, representing millions of legally cast ballots. Congress, controlled by Trump's party, would then install Trump for a second term. The problem with the plan, of course, was that the vice president did not have this authority. When the time came, the vice president refused to comply with Eastman's plan, earning the ire of Trump and his followers and ending the attempt to overthrow the election. On January 20, 2021, Biden was sworn in as president.

A lack of truthful information

During the pandemic, the Capitol insurrection provided the most prominent example of the impact of disinformation. While **misinformation** refers to false or inaccurate information, given sincerely or not, the act of providing disinformation, a special case, involves the deliberate attempt to deceive. As Richard Engel (2019), former Under Secretary of State for Public Diplomacy in the United States, argues, the "rise in disinformation—often accompanied in authoritarian states by crackdowns on free speech—is a threat to democracy at home and abroad. More than any other system, democracies depend on the free flow of information and open debate." Trump's false narrative that "this election was stolen from you, from me, and from the country," in the presence of evidence to the contrary, serves as an example of disinformation (Blake, 2021). A climate of disinformation demonstrated "how seemingly average citizens—duped by a political lie, goaded by their leaders and swept up in a frenzied throng—can unite in breathtaking acts of brutality" (Barry et al., 2021). Trump's insurrectionists tried to alter the course of a valid election.

Historical perspective

In 1887, in the United States, the Electoral Count Act became law. The act established procedures for counting and certifying electoral college votes in both the states and Congress. But the law allows a single member of the House of Representatives or the Senate to object to a state's vote count, which happened on the day of certification, January 6, 2021.

> A small minority of legal scholars have argued that key parts of the Electoral Count Act are unconstitutional, which was the basis of Mr. Eastman's claim that Mr. Pence could simply disregard the law and summarily reject electors of certain key battleground states.
>
> *(Editorial, 2021)*

But nothing in federal law or the Constitution provides the vice president with this authority. The vice president's job on the day of certification is to open the envelopes with the vote totals and read the results, nothing more.

Amplification of discord

The concept of disinformation provides context for this chapter. According to Richard Horton (2020), editor of the medical journal *The Lancet*, "Those who propagate disinformation seek to amplify discord within societies." Trump's lies about the stolen election serve as a prominent example. But, during the pandemic, another example existed: by propagating disinformation, some leaders minimized the threat of the novel coronavirus. They intended to enhance their political position, often for the purpose of reelection. The inability of some leaders to speak the truth—including Trump—reflected the fragilities of modern science-based societies. But the propagation of disinformation also reflected something worse: the failure of democracies to elect honest leaders.

Chapter thesis and organization

The chapter's thesis is that an **epistemic crisis**—a crisis related to knowledge and its degree of validation—constituted an important problem in the era of cascading crises. To develop this thesis, the chapter discusses **epistemology** (the study of the nature of knowledge), infodemic, epistemic injustice, intersectionality, evolving concepts, methods to address the problem, and lessons.

Epistemology

Epistemology addresses how knowledge relates to our worldview and what we believe to be true. As a field of study, it distinguishes between justified belief and opinion. Justified belief is a belief that a person is entitled to hold with respect to a standard of evaluation. Opinion is a judgment or view about an idea or position, not necessarily based on fact. The difference between justified belief and opinion has important implications with respect to the pandemic death toll, social principles, confirmation bias, and epistemic crisis.

Pandemic death toll

Two years into the pandemic, new infections spread throughout society. In developing countries, the distribution of vaccines occurred in inequitable patterns. In developed countries, despite the availability of vaccines, many individuals avoided inoculation. In some regions, the outcome was severe. When the United States reached 700,000 deaths from Covid-19, the crisis became the deadliest pandemic in the country's history. The previous record-holder, the influenza pandemic of 1918–1920, killed 675,000 people.

Vaccination status

According to a *New York Times* analysis of Covid-19, in the United States, during the second year of the pandemic, correlation existed between the death toll and vaccination status:

> An overwhelming majority of Americans who have died in recent months, a period in which the country has offered broad access to shots, were unvaccinated. The United States has had one of the highest recent death rates of any country with an ample supply of vaccines. . . . The deaths are distinct from those in previous chapters of the pandemic...concentrated in the

South, a region that has lagged in vaccinations; many of the deaths were reported in Florida, Mississippi, Louisiana and Arkansas.

(Bosman and Leatherby, 2021)

During this period, the Centers for Disease Control and Prevention (CDC) documented a death rate in the United States for vaccinated individuals equal to 3 percent. The unvaccinated, however, were ten times more likely to die from Covid-19 (Bosman and Leatherby, 2021). Preventing severe illness and death, vaccine mandates were effective. As Zeynep Tufekci (2021) argues, "Science's ability to understand our cells and airways cannot save us if we don't also understand our society and how we can be led astray." Why did some individuals maintain their unvaccinated status? Were the reasons examples of social dysfunction? The following section addresses these questions.

Reasons for noncompliance

First, propaganda in online communities contributed to an erosion of trust. According to a study in *Scientific American*, "misinformation spread by elements of the media, by public leaders and by individuals with large social media platforms . . . contributed to a disproportionately large share of Covid-19 burden" (Bagherpour and Nouri, 2020). In some countries, a partisan divide existed between anti-science, anti-government, and anti-vaccination members of a population on the one hand and pro-science, pro-government, and pro-vaccination members on the other. However, despite political partisanship, many individuals chose all available methods to secure their personal health, including vaccinations. For example, after vaccines became available, individuals over 65 received shots at high rates. As a result, "misinformation is not destiny" (Tufekci, 2021). The problem is that "spreading lies has never been easier. On social media, there are no barriers to entry and there are no gatekeepers. . . . It's far easier to create confusion than clarity" (Engel, 2019).

Second, racialized health inequities, with historical roots in the provision of medical care, existed as a complicating factor. Because the health system may mistreat and provide inferior care for individuals with minority status, they may hesitate to seek vaccinations.

Third, vaccine hesitancy perpetuated uncooperative behavior. These individuals often claimed they were "doing more research" and trying to maintain their dignity (Tufekci, 2021).

Finally, some were "confused and concerned, rather than absolutely opposed to vaccines" (Tufekci, 2021). Those who were vaccine-willing but unclear about potential outcomes existed alongside those who were vaccine-resistant. Some of the unvaccinated worried about previous conditions or side effects. Others were uninformed or lacked advice from medical professionals. Many were scared of needles or concerned about long-term health effects. They exhibited less trust in healthcare institutions. In the United States, many lacked health insurance.

Together, these reasons served as a roadblock to recovery: "Responding to . . . societal dysfunctions has been among the greatest challenges of this pandemic, especially since this includes a political and media establishment stirring up resentment and suspicion to hold on to power and attention" (Tufekci, 2021).

Social principles

The institution of public health represents "the commitment of people living in a society (past and resent) to the twin ideas of **solidarity** and collective action" (Horton, 2020). Solidarity—the feelings of empathy and responsibility that people share—stands in opposition to the principles of competition and individualism. Collective action occurs when people work together for a common purpose. The creation of and support for public health suggests that society is prepared to allocate resources to the institutions that protect and strengthen human lives. But the willingness to act for others depends on justified belief, the basis of functional society. The problem is that a pandemic tests justified belief, especially with respect to a standard of evaluation.

Confirmation bias

The disease outbreaks SARS-Cov-1 in 2003, MERS in 2012, and Ebola in 2014 came from animal hosts. However, with SARS-Cov-2 in 2019, warnings went unheeded. In some countries, such as the United Kingdom and the United States, the reality of infectiousness of the novel coronavirus did not initially provide the impetus for resource mobilization. Compliance with nonpharmaceutical interventions was insufficient. In this environment, why did disinformation persist? Some individuals were "guilty of confirmation bias—ignoring information that doesn't match (their) own view or experience in the world" (Horton, 2020). They sought information that confirmed their beliefs:

> Disinformation sticks because it fits into our mental map of how the world works. The Internet is the greatest delivery system for confirmation bias in history. The analytical and behavioral tools of the web are built to give us the information we agree with.
>
> *(Stengel, 2019)*

Epistemic crisis

These examples demonstrate an epistemic crisis, entailing a state of affairs when **partisans**—strong supporters of a party, cause, or person—disagree on what constitutes fact. First, partisans may doubt whether they "share realities with others and feel an enormous strain in being confronted with multiple and rapid-fire views of news, politics, cultural circumstances, planetary dangers, and economic and self-interests" (Rowell and Call-Cummings, 2020). Second, social media sowing seeds of disinformation exist alongside collaborative systems of technology, cross-cultural empathies, and democratic principles. Third, an "information war" with public sectors and non-state actors "creating and spreading narratives that have nothing to do with reality" complicates public discourse:

> The players in this conflict are assisted by the big social media platforms, which benefit just as much from the sharing of content that is false as content that is true. Popularity is the measure they care about, not accuracy or truthfulness.
>
> *(Stengel, 2019)*

Infodemic

During the coronavirus pandemic, the World Health Organization (WHO) used the term "infodemic" to describe the "overflow of information, some of it true and some of it not, which hampers a reliable and effective response" (Horton, 2020). The WHO became so concerned about the infodemic that it created a new unit, the Information Network for Epidemics, to counter the effects of misinformation. "The crisis (presented) an opportunity to bring the world together drawing upon shared understandings to drive common effort, but forces both political and grassroots (were) acting otherwise" (Ramos and Nycyk, 2020).

Internet and social media

The Internet and social media amplified the infodemic. Originally conceived as places of collaboration and sharing, for many users these platforms transitioned into combative and mistrusted crowded spaces, instruments of "harm used by various actors to control and influence our lives" (Ramos and Nycyk, 2020). Cyberbullying, hacking, and scamming persist. "Fake news, alternative facts and viral conspiracy theories, a conjunction between cynical demagoguery and cultural views and standpoints, have amplified existing fissures and conflicts, and the contradictions we face in creating a sane future" (Ramos and Nycyk, 2020). These problems challenge the concept of truthfulness, fracture the public sphere, and undermine the fabric of society. Reasons include epistemological fracturing, the flow of information, and protected space.

Epistemological fracturing

The Internet and social media create epistemological fracturing when groups inhabit mutually exclusive worlds. Through these forms of technology, some individuals perpetuate conspiracy and polarization, refusing "rational, thoughtful and organized debate without resorting to tribalism or political leanings, abusive trolling or suppressing any side's views by shutting down their side of the debate" (Ramos and Nycyk, 2020).

Flow of information

Writing in *Nature Physics*, Mark Buchanan (2020) argues that, on social media, we underestimate the flow of information. The outcome: falsehoods spread quicker than facts.

> In the United States, a study early in the coronavirus pandemic found that the volume of low-credibility information about the virus shared on Twitter fully matched the volume of more legitimate news coming from *The New York Times* and the Centers for Disease Control.
>
> *(Buchanan, 2020)*

Protected space

In *The Extended Mind: The Power of Thinking Outside the Brain*, Annie Murphy Paul (2021) argues that physical barriers eliminate distractions, provide privacy, and enhance creativity: "putting oneself on display consumes mental resources, leaving less brainpower for the work itself." An absence of protected space increases **cognitive load**, the amount of used working memory resources. Social media eliminates the barriers that shield us from misinformation, increases cognitive load, and encourages conspiracies, discrimination, and stereotypes. The outcome: "when our minds are otherwise occupied, we resort to mental shortcuts—convenient stereotypes, familiar assumptions, well-worn grooves. These are the thoughts that come most readily to mind, that take the least mental energy to generate" (Paul, 2021).

Fake news

After Donald J. Trump was inaugurated as U.S. President, in 2017, the term "fake news" soared into the public vocabulary: he argued that news that was unsupportive of him was fake. Many of his supporters questioned the efficacy of vaccinations, criticized mask mandates, and disparaged healthcare experts. Why did this dynamic persist? Jennifer Rose (2020) of Queen's University targets negative epistemic postdigital inculcation, "interrelationships between our implicit learning, fake news, and digital media. It occurs when we are repetitiously exposed to fake news enabled by digital media." Trump targeted scientists, health experts, and political opponents. "While the specific terminology of fake news surfaces to discredit truthful news, the neologism (newly coined expression) has stayed within our imaginations worldwide" (Rose, 2020).

Malinformation

As this chapter explains, misinformation occurs when an individual unknowingly spreads false or misleading information. Disinformation occurs when an individual knowingly spreads false or misleading information. But malinformation, commonly associated with fake news, occurs when an individual propagates false information with the intention to cause harm. "Each of these information disorders illuminates the intention behind the spread of misleading, fabricated, misconstrued, impostered, manipulated, false, and inaccurate information" (Rose, 2020).

Susceptibility

What factors increase the susceptibility of individuals to believe in fake news? The psychologists Gordon Pennycook and David Rand (2019) argue that, in periods of political division, "our ability to reason is hijacked by our partisan convictions" (Pennycook and Rand, 2019). Individuals may participate in groupthink, the practice of thinking and making decisions in a way that discourages responsibility while individuals "fail to exercise . . . critical faculties" (Pennycook and Rand, 2019). That is, individuals may demonstrate characteristics of mental laziness. But susceptibility to fake news involves both groupthink and laziness. With groupthink, "people use their intellectual abilities to persuade themselves to believe what they want to be true rather than attempting to actually discover the truth" (Pennycook and Rand, 2019). Moreover, "the main factor explaining the acceptance of fake news could be cognitive laziness, especially in the context of social media, where news items are often skimmed or merely glance at" (Pennycook and Rand, 2019).

Public disinformation

During the pandemic, government leaders often "resorted to political disinformation campaigns in order to defend their own roles in managing the outbreak" (Horton, 2020). They often failed "to meet their epistemic duties by relying upon data, models, and evidence of good quality to justify their actions" (Winsberg et al., 2020). The outcome was objection to methods of intervention. If leaders minimized the importance of masks and social distancing, followers were susceptible to disinformation, refusing reason and informed judgments. "Just as there has been a struggle to contain the outbreak, there is a struggle to control the way the public views government management of the outbreak" (Horton, 2020).

Liberty

In democratic societies, liberty exists as a fundamental value. Restrictions should occur in exceptional cases:

> Basic liberties can be restricted only if justifications survive strict scrutiny, while restrictions on non-basic liberties still require significant

justifications. The stronger the imposition and the greater the potential harm it imposes, the stronger the needed justification.

(Winsberg et al., 2020)

A global pandemic meets the justification of strict scrutiny. To slow the spread of a deadly virus, society must restrict human activity. For government, meeting an epistemic duty means relying on accurate information and conveying it to the public. But, during the pandemic, some public officials failed to meet this threshold, creating a crisis of governance.

Epistemic injustice

Miranda Fricker (2007), in her influential book, *Epistemic Injustice: Power and the Ethics of Knowing*, describes the existence of a "distinctively epistemic kind of injustice." When prejudice causes an individual to deflate the level of credibility of a speaker, or when a gap in collective interpretive resources causes an individual to lack the wherewithal to understand social experiences, epistemic injustice occurs. The first example exists when a police officer doubts the credibility of a marginalized member of society. The second example occurs when an individual believes a leader who perpetuates disinformation. In her book, Fricker (2007) elucidates two ethical outcomes, conveying knowledge and understanding experiences: "Since the ethical features in question result from the operation of social power in epistemic interactions, to reveal them is also to expose a politics of epistemic practice."

Types of epistemic injustice

According to Benjamin Sherman and Stacey Goguen (2019), editors of *Overcoming Epistemic Injustice: Social and Psychological Perspectives*, four types of epistemic injustice exist (Table 7.1). Although the types do not constitute an exhaustive list, they cover the examples in this chapter.

TABLE 7.1 Types of epistemic injustice

Type	Meaning
Testimonial injustice	Using prejudice to assign less credibility to others
Hermeneutical injustice	When a gap in collective interpretation creates disadvantage
Epistemologies of ignorance	When ignorance creates a context for misinterpretation
Willful hermeneutical ignorance	Restricting the ability of others to understand the truth

Source: Sherman and Goguen (2019).

Testimonial injustice

When prejudice—a preconceived opinion that is not based on reason or experience—reduces the credibility of others, testimonial injustice occurs. Prejudice may result from a stereotype, the oversimplification of individual characteristics. Those with prejudice may not trust others because of who they are, believe that members of other groups should hold leadership positions, or trust pronouncements of an opposing party. But layers of complexity exist. First, testimonial injustice is incorrect, but part of the act may be ethically culpable, such as the potential for harm. Second, testimonial injustice may include **systematic prejudice** when prejudice is commonly practiced. Third, testimonial injustice may involve deflated credibility judgments, but not inflated credibility judgments. Finally, life experiences may create stereotypes that present a misleading body of evidence about certain groups, leading to testimonial injustice (Sherman and Goguen, 2019).

Hermeneutical injustice

The concept of hermeneutics refers to the theory and methodology of interpretation. Hermeneutical injustice exists as a collective problem. It requires collective solutions. As Miranda Fricker (2007) explains, hermeneutical injustice occurs "when a gap in collective interpretive resources puts someone at an unfair disadvantage when it comes to making sense of their social experiences." As an example, for generations women struggled with sexual harassment but lacked the legal support to address the problem, especially when it fell short of physical assault. Sexual harassment was often interpreted as friendly or complimentary, even when it was harmful and unwelcome. But the introduction of the term "sexual harassment" in the 1970s into the legal environment led to the identification of sexual harassment as harmful and illegal behavior. This change enabled women to both describe their grievances and fight for reform.

Not all gaps in collective interpretive resources are unjust. The gaps may result from discoveries, novel situations, or technological advances. The consequence is that it may take time for a shared conceptual framework to emerge. With sexual harassment, an unjust practice, the fact that it was common but not addressed pointed to a larger problem. Women were excluded from leadership roles, including policymaking and political office, participation necessary for shared meaning and understanding. "Many people collectively would need to be aware of certain experiences and have a shared way of understanding them—a collective change was needed" (Sherman and Goguen, 2019). Hermeneutical injustice challenges those with privilege to identify gaps in their understanding. When marginalized groups do not participate in collective decision making, their voices go unheard, and hermeneutical injustice persists.

Epistemologies of ignorance

Epistemologies of ignorance are distinct from both testimonial and hermeneutical injustice. First, ignorance may be innocent, resulting from a lack of experience with diversity and difference. In sheltered environments, it may be difficult or impossible to acquire knowledge. Second, ignorance may not be innocent, stemming from misguided but purposeful beliefs. With privilege, one group experiences a relatively favorable socioeconomic position. For those with privilege, collective understanding exists. This position generates a benefit: agreeing to a certain belief means acceptance into the privileged group. But, with privilege, collective understanding may involve distortions of reality. The privileged may misinterpret the world to maintain their misguided thinking of superiority (Sherman and Goguen, 2019).

Willful hermeneutical ignorance

Willful hermeneutical ignorance combines elements of willful ignorance and hermeneutical injustice. It occurs when influencers deny others "the opportunity to develop the conceptual resources needed to make sense of their own experience" (Sherman and Goguen, 2019). An example is when sexist discrimination denies women the opportunity to participate fully in the economy. This toxic belief marginalizes a substantial portion of the working-age population and prolongs necessary changes. Willful hermeneutical ignorance means that the privileged may disparage or ignore a new way of thinking. "In such cases, the privileged take part in creating situations where they are unable to understand the experiences of the marginalized, through earlier patterns of derision, neglect, or uncharitable responses" (Sherman and Goguen, 2019). The privileged decide not to think about or understand novel concepts, such as when women participate in the labor force the entire economy benefits.

Toxicity and danger

Fractures in contemporary society, including the four types of epistemic injustice, create toxicity and danger, evident in the epistemological divide. From the contested nature of the Internet, political environment, and social media, the problems create incompatible perspectives, evident in the speed of misinformation, distrust of science, and advancement of knowledge.

Speed of misinformation

In science, researchers establish questions, derive hypotheses, formulate predictions, make observations, and create experiments to test the hypotheses. The process requires time, effort, and resources. But the spread of misinformation occurs quickly. In a relative context, the process of scientific advance challenges an individual's capacity to focus. But, in the frenzy of Internet exchange, individuals may lose the patience to process nuanced but important information. Information overload discourages careful reflection and impairs civic competence:

> people give up on making informed decisions in the face of competing viewpoints and simply adhere to their preexisting biases. There are now around 6,000 tweets per second sent out on Twitter, 23 billion text messages sent daily around the world, and equally staggering numbers of posts made to Facebook, YouTube, Instagram, Weibo, Reddit and other social network platforms.
>
> *(Rowell and Call-Cummings, 2020)*

Distrust of science

A distrust of science stems from several sources: the existence of educated elites, mainstream media, and politicians of opposing parties. "This orientation seeks to replace evidence and thoughtful analysis with bile and the group hysteria of mass rallies and political propaganda" (Rowell and Call-Cummings, 2020). Many leaders, such as Donald J. Trump, practice principles of propaganda, including the creation of the illusion of truth. An example of his is the propagation of unproven cures to a disease, such as bleach (Eaton et al., 2020; Frenkel and Alba, 2020). This disinformation challenges the informed citizen to reflect on the role of truth and democratic ideals in a well-functioning society.

Advancement of knowledge

The concept of epistemicide, the destruction of existing knowledge, is especially relevant for indigenous and underrepresented groups. Because of the dominance of preeminent technology platforms, a myopic view of the production of knowledge may persist across geopolitical contexts. Epistemologies of the marginalized, especially those of intersectional status, are crowded out (Rowell and Call-Cummings, 2020).

Case study 7.1 Digital environments and epistemic injustice

By telling fabricated stories, propagators of disinformation harm listeners in their capacities as knowledgeable individuals. Digital environments accelerate this trend, depriving those who are vulnerable to disinformation of their epistemic agency, their ability to act independently or make rational choices. To establish community and share information, online groups and social media establish an assemblage of big data on their attitudes, behaviors, and intentions. The problem is that the data may be exploited. Examples include raising money for political campaigns, arguing for social positions, or propagating false beliefs.

The gathering of data about behavior and identity establishes power asymmetries. In digital environments, those who capitalize from online platforms may purport to better understand the individuals who consume online information than the individuals themselves. To maintain membership in a group, the thinking goes, the individuals must contribute to the group, argue for a position, or propagate an untruth, even in the presence of evidence that this behavior runs counter to the collective good, personal health, or financial stability. That is, the process of manipulation weakens individual agency.

These examples undermine confidence about knowledge, especially when knowledge is considered to be infallible. The targets of epistemic injustice may think their candidate won the election, despite the vote count that concludes otherwise. They may think a false accusation about a member of an opposing political party exists as truth, despite the absurd nature of the accusation. They may think a specific social position establishes a safe community when the position creates more instability. With online content, an endless potential for injustice exists.

As Gloria Origgi and Serena Ciranna (2019) explain, individuals susceptible to disinformation are "diminished as knowers, especially in the most intimate part of (their) epistemic competence." The capacity to reason entails multiple forms of identity, including the capacity to provide knowledge. Practicing testimonial injustice by providing disinformation exists as an equivalent to wronging others as human beings. In this process, who suffers? The

> receiver of information, due to her biases and prejudices, will end up with less information than she would have had if she had considered the speaker at her face value instead of applying biased filters to her credibility assessment. Thus, in a sense, the hearer inflicts on herself an epistemic offense.
>
> (Origgi and Ciranna, 2019)

A lack of social knowledge impacts society as a whole. Society's immersion in digital environments accelerates this trend. For epistemology, a technologically-advanced society presents problems. How do members of partisan groups seek online information? How does communication occur? With respect to online searches and monitoring, what is the role of trust? The digital environment, with its evolving methods of interaction, creates the potential for epistemic injustice, "because the representation and the selection of the raw data gathered in order to be intelligible to the users is done through algorithmic procedures that are determined by the owners of the platforms according to their specific interests" (Origgi and Ciranna, 2019). The results: users do not receive the information they need; biased representations of consumers of online content persist; big data establishes an informational prejudice against providers of information when they lack the ability to provide truthful or relevant content; and individuals who propagate disinformation serve as unsuitable participants in public discourse. Across multiple platforms, the proliferation of the digital environment perpetuates epistemic injustice.

Intersectionality

Patricia Collins (2019), the Distinguished University Professor of Sociology Emerita at the University of Maryland, College Park, explains the importance of intersectionality: "race, class, gender, sexuality, ethnicity, nation, ability and age operate not as unitary, mutually exclusive entities, but rather as reciprocally constructing phenomena that in turn shape complex social inequalities." She applies the concept to **epistemic oppression**—exclusion that reduces contributions to knowledge production—as a defining feature of intersecting systems of power and influence. The concept of epistemic oppression extends the analysis beyond social groups, revealing actions that perpetuate inequality. It also encourages the use of epistemological frameworks. Applications include identity politics, epistemic agency, and resistance.

Identity politics

Those who benefit from the existing order may prefer inertia to activism, demonstration, and mobilization. When these forces apply pressure, those who benefit may ignore intersectional frames, discredit those seeking new social arrangements, and resist progressive change. Examples include opposition to the Black Lives Matter (BLM) and environmental justice movements. For those who experience multiple layers of oppression, identity politics facilitates alliances. But the idea exceeds self-preservation. Identity politics influence but do not reduce characteristics of identity—the fact of being who or what a person is—to subjective notions of the individual. In the presence of intersectionality, individuals are empowered by epistemic practices (Collins, 2019).

Epistemic agency

In communities that fight for social justice, individuals should experience testimonial recognition: "They should participate equitably albeit differently in intersectionality's knowledge production and, if required, enjoy access to epistemic authority to shape intersectionality's definitions" (Collins, 2019). But, in social media, those who propagate fake news, even with countervailing evidence, may silence the voices of those who are skeptical or support propagation of the truth. Those who disavow disinformation should experience epistemic agency: control over their beliefs. The reason: "Identity politics claims the authority of one's own experiences" (Collins, 2019).

Resistance

Subordination, epistemic oppression, and identity politics underscore the need for empowerment. One view is that society requires less social activism, the way a movement away from social characterizations such as class and gender leads to less emphasis on discrimination and oppression. Another view is that the persistence of epistemic oppression, identity politics, and epistemic agency requires more social activism. Intersectional characteristics such as race, class, and gender elevate inequities within society, such as environmental injustice, income inequality, and the uneven effects of climate change. Do social agents who emphasize intersectional outcomes have an obligation to resist inequities? Jose Medina (2013), the Walter Dill Scott Professor of Philosophy at Northwestern University, argues the answer is yes, because "those who live under conditions of oppression—however they happen to inhabit contexts of domination (as victim, as a bystander, as both victim and oppressor, etc.)—have an obligation to resist."

Evolving concepts

Miranda Fricker (2019) argues that the reason to fight epistemic injustice is to "delineate a distinctive class of wrongs, namely those in which someone is ingenuously downgraded and/or disadvantaged." First, prejudice leads to misjudgment and inferior treatment of others as epistemically unequal. Second, this behavior creates negative economic, political, and social environments. Third, epistemic injustice perpetuates inequality, reflecting the intersectional nature of the problem. Those perpetuating epistemic injustice may not recognize the negative consequences of their attitudes, behaviors, and beliefs. It is, therefore, important to consider discriminatory epistemic injustice, occurring when individuals receive "less than their fair share of an epistemic good, such as education, or access to expert advice or information" (Fricker, 2019). Individuals may be wronged in their capacities as epistemic subjects. In this and other chapters, many examples exist. During a pandemic, leaders who propagate misinformation mislead their followers. With racial injustice, negative stereotypes prevent an understanding of the problem. With domestic violence, those who limit the crisis to personal matters ignore structural sexism and misogyny—the dislike of, contempt for, or ingrained prejudice against women—and why domestic violence persists across cultures and eras.

Trust and distrust

Katherine Hawley (2019), Professor of Philosophy at the University of St. Andrews, extends the analysis, arguing that overlapping concepts of trust and distrust exist. Trust exists as a personal leap into the unknown or a pragmatic choice, focusing on costs and benefits. Either way, trust exists in relationships, digital space, and public interactions. We offer varying degrees of trust for leaders, individual actions, and behaviors. At the same time, the absence of trust—distrust—exists as a multilayered concept. First, we may be unsure about a potential outcome. If an individual does not have training in a particular field, we may not trust or distrust the individual to contribute. Second, without evidence, it is difficult to determine the appropriateness of trust or distrust. If one needs someone to tend a garden, it is difficult to trust or distrust a neighbor with the responsibility if the neighbor does not have a garden or express any interest in gardening. Third, trust may be difficult to express. A leader may promise to act in the interest of others but possess a record of dishonesty. Together, the examples demonstrate that it is appropriate and useful to consider the interconnected nature of trust and distrust, especially with confidence and testimony.

Confidence

Trust involves practical situations. We rely on others to act in specific ways. But trust entails confidence that others will carry through on tasks or make competent pronouncements. It is, therefore, important to distinguish between low expectations and subjective beliefs. An individual may distrust politicians but possess low expectations about a local restaurant's service. "Distrust embodies a moral criticism involving attitudes such as resentment and may have a distinctive emotional color" (Hawley, 2019). Low expectations, in contrast, lack those features, existing as a practical accommodation in an imperfect world, not a judgment about moral depravity.

Testimony

Testimony entails the methods in which individuals provide information to others. Conversations, posts, and speeches provide information. But, with these forms of testimony, observers may or may not believe what they hear or read. Contextual factors, such as the political environment and social media, interact to influence the level of credibility assigned to the provider of information. If the provider is the leader of an opposing party, a listener may assign zero credibility to the provider's pronouncements. In effect, individuals assess a speaker in terms of not only competence, goodwill, intentions, and sincerity, but also the ability to act (Hawley, 2019). Individuals may judge a leader, for example, as untrustworthy but skilled. To persuade followers, a leader may declare it is important to "storm the Capitol," and "I will be with you." The first statement may achieve the desired result: followers storm the Capitol. The second pronouncement is insincere: the leader has no intention of joining the group but makes the statement anyway.

Addressing epistemic injustice

To fight epistemic injustice, Nadya Vasilyeva and Saray Ayala-Lopez (2019) propose a structural framework. Epistemic injustice occurs when individuals establish personal representations of social categories. They may practice stereotypes, such as believing one group is naturally more talented or gifted than another group. Cultural histories and norms may encourage the perpetuation of stereotype. The problem is that those who practice stereotypes act on their belief, belittling or disrespecting members of other groups. But structural reform alters the environment that shapes and encourages mistaken attitudes, beliefs, and behavior:

> instead of trying to upgrade our flawed minds, full of problematic associations between social groups and traits, we could turn to the correlations that exist in our corrupted society and are picked up by our minds as we form representations useful for navigating the world we inhabit.
> *(Vasilyeva and Ayala-Lopez, 2019)*

This book addresses patterns of income inequality (Chapter 3), racial injustice (Chapter 5), and domestic violence (Chapter 6). A common theme involves unjust structural dynamics. Relevant to this chapter, society's means of production and systems of representation create a structure of epistemic injustice. This reality establishes a locus of intervention: the social dynamics that maintain discrimination and injustice. Instead of addressing the deficiencies of understanding in the minds of individuals, society should alter the context of mistaken beliefs: "Once the problematic associations disappear from the environment, relying on mental shortcuts to reason about it would not lead to discrimination and injustice" (Vasilyeva and Ayala-Lopez, 2019). That is, society should reduce income inequality, eliminate racial injustice, and prevent domestic violence. With this structural approach, two benefits exist. First, the approach addresses ongoing problems. Second, the approach does not fight characteristics of human psychology. Instead, structural intervention alters the social system in ways that align behavior with evidence.

Cultivation

Alex Madva (2019), Professor of Philosophy at Cal Poly Pomona, argues that, to overcome epistemic injustice, individuals should cultivate testimonial and interpretive virtues. A **virtue** is a behavior showing high moral standards. Cultivating a virtue means exhibiting intentionality in doing what is right and avoiding what is wrong. In particular, "the pivot away from epistemic injustice depends in part on moments of self-critical awareness, states of cognitive dissonance in which an individual realizes that she may, for example, be underestimating an interlocutor's credibility due to stereotypes or prejudices" (Madva, 2019).

This emphasis on cognition includes the assumption that humans have a tendency toward stubbornness, either hardwired from birth, socially reinforced, or both. As a result, epistemic-virtue cultivation involves feelings, perception, and thought. Learning new ways to process and apply information reduces or eliminates the tendency to focus on prejudice and stereotype. Instead, individuals self-reflect, cultivate epistemic habits, and neutralize the effects of stereotypes or prejudices. The result, epistemic-virtue cultivation, entails the development of humility and open-mindedness.

Two challenges exist. First, an individual's first-order epistemic intuition may be incorrect. Second, an interactional dimension of epistemic virtue involves the evaluation of a speaker's ideas and intentions. Madva (2019) asks: what is the cognitive architecture of an epistemically unbiased mind? Although humility and open-mindedness play important roles, a multidirectional orientation exists. A credible evaluation of a speaker's pronouncements involves three areas: virtue cultivation, the degree of intentionality of the speaker, and structural intervention. Individuals have control over the first, the ability to be mindful of the second, and the possibility of influencing the third. In this framework, the intentionality of the speaker claims an important position. According to Madva (2019):

> Epistemic virtue requires being the sort of person who reliably and responsibly supports institutions that generate, disseminate, and retain knowledge. This involves taking steps to promote (or, minimally, not taking steps to impede) the creation, revision, and maintenance of just bodies of knowledge. It also requires attention to the ways that individuals' social locations and situations inform their beliefs and other epistemic dispositions.

That is, individuals should cultivate the cognitive architecture of the epistemically open mind. To this end, they should practice humility and open-mindedness with respect to personal thoughts and attitudes, display courage and honesty in interactions with others, and support institutions that provide accurate information.

Patience, persistence, and critical reflection

In the presence of epistemological fractures, it is important to envision an inclusive and transparent future that encourages informed views, places respect at the center of discourse, and builds knowledge democracies. To move from the contemporary climate of Internet toxicity and information overload to this idealized future, societies must exhibit patience, persistence, and critical reflection. Amir Bagherpour and Ali Nouri (2020) recommend several interventions that address the problem of misinformation, including cooperation between social media companies and governments, detection of misinformation, matching public health responses and government capabilities, public messaging, and support for science.

Cooperation

To reduce the flow of misinformation, social media companies should work in tandem with government and public health officials. On social media platforms, an important category of misinformation entails the mischaracterization of messages from public officials. During the coronavirus pandemic, while many social media platforms attempted to address the problem, their actions were delayed, reactive, and drowned out by fresh news cycles. This inefficiency led to the circulation of misinformation to unwitting consumers. For this reason, a robust partnership between social media companies, public health officials, and government must "identify common sources of misinformation; proactively anticipate future misinformation from those sources; and enable its removal in a near real-time fashion" (Bagherpour and Nouri, 2020). To establish credibility, the process must be consistent, nonpartisan, and transparent.

Detection

The detection of misinformation and propagation of accurate information requires the disciplines of behavioral analytics and data science. These fields statistically analyze the relationship between variables and provide visualization of the results. In public health, individuals who communicate trends on disease outbreaks should be fluent in these techniques. Understanding the preferences of those who operate on social media platforms requires the delivery of salient information. Society should target these audiences, incentivizing individuals to undertake preferred behaviors (Bagherpour and Nouri, 2020).

Matching

The delivery of accurate and trustworthy information should match the advice of public health officials. The spread of misinformation provides an obstacle to this effort. For example, information and guidance should accompany testing for infectious diseases. Clinics should justify vaccination mandates. A sufficient supply of masks should complement guidance for wearing masks. Information campaigns should accompany nonpharmaceutical interventions. To flatten the epidemic curve and fight the spread of disease, a comprehensive policy framework should fight an infodemic (Bagherpour and Nouri, 2020).

Messaging

In addition to cooperation, the detection of misinformation, and matching public health responses with government capabilities, a dynamic public messaging campaign should pass a review process, enabling health officials to interact with social media consumers in real time. This increases the potential to debunk misinformation. But, to increase the effectiveness of messaging, public health officials should first think like media consumers. Their messages should be timely. They should speak directly to the audience. They should take advantage of the platforms that are popular with younger generations:

> Dynamic conversations and proactive messaging between public health officials and the public can be more impactful than removing false information from social media platforms, especially since removal typically occurs long after a significant number of individuals have already been exposed to the false message.
>
> *(Bagherpour and Nouri, 2020)*

Support

To solve a crisis, society requires leaders who support the scientific process. In contrast, leaders who provide false information for political gain serve as a roadblock. On social media, the pronouncements of leaders occupy a prominent place in public discourse. As a result, society should support social media platforms when they disseminate accurate and truthful information on public policies. A coordinated campaign of science and public health should combine the capabilities of influencers from different social media platforms, amplifying the provision of consistent and factual messages across media outlets (Bagherpour and Nouri, 2020).

Lessons

Address indignities

The outcomes of epistemic injustice, including false beliefs, ignorance, and silencing, create layers of misconception. At the local level, marginalization from hermeneutical practices and the loss of agency from epistemic ignorance prevent the dissemination of truthful information. At the national level, structural blockages in the sharing of knowledge and the deliberate propagation of disinformation create suffering. These problems result from the systematic nature of socio-epistemic practices. But epistemic injustice includes more than harmful outcomes. According to Matthew Congdon (2019), Professor of Philosophy at Vanderbilt University, there is a "deeper intrinsic indignity at work in various forms of epistemic injustice, as in, for example, unjustly downgrading one's trustworthiness via testimonial injustice or the forms of racial domination." The notions of dignity, personhood, and virtue may address these wrongs, especially for oppressed individuals. Epistemic justice includes virtues (accuracy and sincerity) for speakers and virtues (humility and openness) for listeners (Congdon, 2019).

Create agency

To reduce epistemic injustice, society should distinguish between the historical conditions that impede agency and the accepted conditions that facilitate agency.

The conditions that establish agency may be limited, prevented, or restricted by systems and material structures of injustice. For those who perpetuate disinformation, it is important to understand their choices. According to Lorenzo Simpson (2019), Professor of Philosophy at Stony Brook University,

> Understanding such choices . . . can bring into focus cases where the agent making the choices is not culpable for their failure to make life-enhancing choices (while, of course, still leaving open the possibility that in some cases they may be), but where instead an injustice of an epistemic sort is present.

In this case, compound injustice refers to the act of failing to consider the structural environment that compromise the individual's ability to act. While blamed or judged for a lack of agency, an epistemic limitation—stemming from commonly-held beliefs, the need to maintain status in a group, or ignorance—compromises ethical and truthful decisions (Simpson, 2019).

Establish universals

A factual position may not correspond to expectations. As an example, because a leader claims he will win an election, a supporter may expect this eventuality. The individual may not believe an alternative outcome, such as the loss of an election. If a loss occurs, the individual may not alter personal conviction. The individual may not fathom a world in which the leader loses. Expectations, therefore, include investment in identity. Susan Babbitt (2019), Associate Professor of Philosophy at Queen's University, argues that

> epistemic freedom is about how to live, morally and politically. It might seem to be principally about thinking on one's own. Yet 'one's own' thinking needs discovery. And such discovery depends upon conditions, which often need to be brought about through moral, political and even personal action.

But both actions and thinking, Babbitt continues, depend on **universals**, concepts of general application. An example is the concept of freedom. An individual's thinking may be private, but universals are social, shared, and generally accepted. For a functional society, universals must unify experiences. The reason: knowledge entails judgments about universals. Any act, moral or nonmoral, social or private, is a function of universals, which establish what we think, how we deliberate, and what we do. But universals such as freedom may be more grounded than others, depending on cultural, historical, political, and social conditions (Babbitt, 2019). For a group, freedom may mean life without government mandates, even if this choice increases the risk of illness or death. A group with a different political predilection may disagree, viewing freedom as safety from disease. Epistemic injustice, therefore, calls into question whether a society may establish universals.

Strengthen communities

Epistemic agency addresses how marginalized communities resist multiple forms of oppression. Nancy McHugh (2019), Professor of Philosophy at Wittenberg University, argues that the "epistemic agency of communities . . . (raises) critical questions about how knowledge is constituted, who counts as a legitimate knower, and how conditions of epistemic injustice shape communities and individuals." The idea is that, because individuals operate in a cultural context, communities create knowledge and serve as agents of epistemology. Through coalitions and instruction, individuals learn the attitudes, beliefs, and ideas of communities, especially how to view opportunity, success, discrimination, and oppression. Communities teach individuals how to think about economic, political, and social contexts, including how a leader, police officer, or other influential member of society should act. Even more, communities convey

> a set of practices that can livingly confer tradition, but also as practices that confer habits of privilege, experiences of marginalization, ways of viewing our own and others' bodies, practices that sediment social relationships and interactions, and an epistemic lens through which to experience and know the world.
>
> *(McHugh, 2019)*

As a result, communities that fight racial and environmental injustice, such as indigenous groups fighting the implementation of a pipeline on their land, act through social experiences.

Summary

During the pandemic, the U.S. Capitol insurrection provided an example of the impact of disinformation. But, because a fair and transparent election led to a decisive result, Joseph P. Biden became president. While misinformation refers to false or inaccurate information, given sincerely or not, the act of providing disinformation, a special case, involves the deliberate attempt to deceive. Epistemology, the theory of knowledge, focuses on methods, scope, and validity, including exclusion, silencing, and the systematic distortion or misrepresentation of meaning. During the coronavirus pandemic, the term "infodemic" was used to describe the flow of disinformation. While the types of epistemic injustice include testimonial injustice, hermeneutical injustice, epistemologies of ignorance, and willful hermeneutical ignorance, an intersectional perspective provides a way to consider how individuals assess the flow of information. Structural intervention and virtue cultivation serve as two methods to address the problem of epistemic injustice. In addressing epistemic injustice, several lessons exist. Society should address indignities, create agency, establish universals, and strengthen communities.

Chapter takeaways

LO1   While misinformation refers to false or inaccurate information, given sincerely or not, the act of providing disinformation, a special case, involves the deliberate attempt to deceive.

LO2   Epistemology, the study of the nature of knowledge, addresses how knowledge relates to our worldview and what we believe to be true.

LO3   During the coronavirus pandemic, the infodemic described an overflow of information, some of it true and some of it not, which hampered a reliable and effective response.

LO4   When prejudice causes an individual to deflate the level of credibility of a speaker, or when a gap in collective interpretive resources causes an individual to lack the wherewithal to understand social experiences, epistemic injustice occurs.

LO5   Epistemic oppression—exclusion that hinders one's contributions to knowledge production—serves as a defining feature of intersecting systems of power and influence.

LO6   An important reason to address the problem of epistemic injustice is to elucidate specific wrongs, especially when individuals are disadvantaged because of their status.

LO7   Different methods address the problem of epistemic injustice, including structural intervention and virtue cultivation.

LO8   In evaluating epistemic injustice, several lessons exist, including the need to address indignities, create agency, establish universals, and strengthen communities.

Key terms

Cognitive load

Disinformation

Epistemic crisis

Epistemic oppression

Epistemology

Misinformation

Partisans

Solidarity

Systematic prejudice

Universals

Virtue

Questions

1   In what ways does the U.S. Capitol insurrection involve disinformation?

2   As it relates to the pandemic, explain the infodemic. What are the roles of the Internet, social media, fake news, and the public sector response?

3   What are the differences between testimonial injustice, hermeneutical injustice, epistemologies of ignorance, and willful hermeneutical ignorance?

4   With respect to epistemic injustice, what are the roles of the speed of misinformation, suspicions of science, and knowledge production?

5   How does an intersectional perspective relate to epistemic oppression: exclusion that hinders one's contributions to knowledge production?
6   Do you agree that the reason to address the problem of epistemic injustice is to delineate a distinctive class of wrongs? Explain.
7   With respect to addressing the problem of epistemic injustice, several approaches exist, including structural intervention and virtue cultivation. Explain these approaches. In the present, which is most applicable? Why?
8   How might addressing indignities, creating agency, establishing universals, and strengthening communities influence the conversation about epistemic injustice?

References

Babbitt, Susan. 2019. "Epistemic and political freedom." In Kidd, Ian (Ed.), *The Routledge Handbook of Epistemic Injustice*. London: Routledge.

Bagherpour, Amir and Nouri, Ali. 2020. "Covid Misinformation is Killing People." *Scientific American*, October 11.

Barry, Dan, Feuer, Alan, and Rosenberg, Matthew. 2021. "90 Seconds of Rage on the Capitol Steps." *The New York Times*, October 17.

Blake, Aaron. 2021. "What Trump Said before His Supporters Stormed the Capitol, Annotated." *Washington Post*, January 11.

Bosman, Julie and Leatherby, Lauren. 2021. "U.S. Coronavirus Death Toll Nears 700,000 Despite Wide Availability of Vaccines." *The New York Times*, October 1.

Brennan Center for Justice. United States. 2020. Web Archive. "It's Official: The Election Was Secure." December 11. https://www.brennancenter.org/our-work/research-reports/its-official-election-was-secure

Buchanan, Mark. 2020. "Managing the infodemic." *Nature Physics*, 16: 894.

Collins, Patricia. 2019. "Intersectionality and epistemic injustice." In Kidd, Ian (Ed.), *The Routledge Handbook of Epistemic Injustice*. London: Routledge.

Congdon, Matthew. 2019. "What's wrong with epistemic injustice? Harm, vice, objectification, recognition." In Kidd, Ian (Ed.), *The Routledge Handbook of Epistemic Injustice*. London: Routledge.

Eaton, Melissa, King, Anne, Delmayne, Emma, and Seigler, Amanda. 2020. "Trump Suggested 'Injecting' Disinfectant to Cure Coronavirus? We're not Surprised." *The New York Times*, April 26.

Editorial. 2021. "Jan. 6 Was Worse Than We Knew." *The New York Times*, October 2.

Engel, Richard. 2019. *Information Wars: How We Lost the Global Battle Against Disinformation and What We Can Do about It*. New York: Atlantic Monthly Press.

Frenkel, Sheera and Alba, Davey. 2020. "Trump's Disinfectant Talk Trips Up Sites' Vows Against Misinformation." *The New York Times*, April 30.

Fricker, Miranda. 2019. "Evolving concepts of epistemic injustice." In Kidd, Ian (Ed.), *The Routledge Handbook of Epistemic Injustice*. London: Routledge.

Fricker, Miranda. 2007. *Epistemic Injustice: Power and the Ethics of Knowing*. New York: Oxford University Press.

Hawley, Catherine. 2019. "Trust, distrust, and epistemic injustice." In Kidd, Ian (Ed.), *The Routledge Handbook of Epistemic Injustice*. London: Routledge.

Horton, Richard. 2020. *The Covid-19 Catastrophe*. Cambridge: Policy Press.

Madva, Alex. 2019. "The inevitability of aiming for virtue." In Sherman, Benjamin and Goguen, Stacey (Eds.), *Overcoming Epistemic Injustice: Social and Psychological Perspectives.* New York: Roman & Littlefield.

McHugh, Nancy. 2019. "Epistemic communities and institutions." In Kidd, Ian (Ed.), *The Routledge Handbook of Epistemic Injustice.* London: Routledge.

Medina, Jose. 2013. *The Epistemology of Resistance: Gender and Racial Oppression, Epistemic Injustice, and Resistant Imaginations.* New York: Oxford University Press.

Origgi, Gloria and Ciranna, Serena. 2019. "Epistemic injustice: The case of digital environments." In Kidd, Ian (Ed.), *The Routledge Handbook of Epistemic Injustice.* London: Routledge.

Paul, Annie. 2021. *The Extended Mind: The Power of Thinking Outside the Brain.* New York: Houghton Mifflin Harcourt.

Pennycook, Gordon and Rand, David. 2021. "Examining false beliefs about voter fraud in the wake of the 2020 Presidential Election." *The Harvard Kennedy School Misinformation Review,* 2(1): 1–19.

Pennycook, Gordon and Rand, David. 2019. "Why Do People Fall for Fake News?" *The New York Times,* January 19.

Ramos, Jose and Nycyk, Michael. 2020. "The Internet, epistemological crisis and the realities of the future: An introduction to this special issue." *Journal of Future Studies,* 24(4): 1–4.

Rose, Jennifer. 2020. "The mortal coil of Covid-19, fake news, and negative epistemic postdigital inculcation." *Postdigital Science and Education,* 2: 812–829.

Rowell, Lonnie and Call-Cummings, Meagan. 2020. "Knowledge democracy, action research, the internet and the epistemic crisis." *Journal of Future Studies,* 24(4): 73–82.

Sherman, Benjamin and Goguen, Stacey. 2019. "Introduction." In Sherman, Benjamin and Goguen, Stacey (Eds.), *Overcoming Epistemic Injustice: Social and Psychological Perspectives.* New York: Roman & Littlefield.

Simpson, Lorenzo. 2019. "Epistemic and political agency." In Kidd, Ian (Ed.), *The Routledge Handbook of Epistemic Injustice.* London: Routledge.

Stengel, Richard. 2019. *Information Wars: How We Lost the Global Battle Against Disinformation & What We Can do About It.* Washington, DC: Atlantic Monthly Press.

Tufekci, Zeynep. 2021. "The Unvaccinated May Not Be Who You Think." *The New York Times,* October 15.

Vasilyeva, Nadya and Ayala-Lopez, Saray. 2019. "Structural thinking and epistemic injustice." In Sherman, Benjamin and Goguen, Stacey (Eds.), *Overcoming Epistemic Injustice: Social and Psychological Perspectives.* New York: Roman & Littlefield.

Winsberg, Eric, Brennan, Jason, and Surprenant, Chris. 2020. "How government leaders violated their epistemic duties during the SARS-CoV-2 crisis." *Kennedy Institute of Ethics Journal,* 30(3/4): 215–242.

PART III

Building future resilience

8

PROGRESS OR COLLAPSE?

Chapter learning objectives

After reading this chapter, you will be able to:

LO1   Evaluate the implications of a change in the existing order.
LO2   Define the concept of complex societies.
LO3   Demonstrate that investment in complex societies leads to diminishing returns.
LO4   Identify the disruptions that alter complex societies.
LO5   Assert that alterations of the existing order entail a range of potential outcomes.
LO6   Discuss factors that constitute human development.
LO7   Explain the complexity continuum.

Chapter outline

Change in the existing order
Complex society arrangements
Investment in complex societies
Disruptions
A range of potential outcomes
Human development
The complexity continuum
Summary

DOI: 10.4324/9781003310075-11

Change in the existing order

A change in the existing order resulting from the coronavirus pandemic, economic contraction, climate catastrophe, and social instability leads to the establishment of a new reality. But whether the disruptions establish a stronger or weaker position in the society at large depends on several factors that are addressed in this chapter. From the outset, it is important to consider the characteristics of the existing order, forms of disruption, and effectiveness of policy responses. Given the destabilizing effects of a series of cascading crises, analyzing potential outcomes serves as a way to understand the contemporary environment. This method also provides a framework to address the process of change, which includes both positive and negative elements.

Characterization of the modern world

Belief systems, networks, and methods of organization characterize the modern world. They constitute elements of the existing order. Societies contain economic, political, and social systems that adapt or contract, rise or fall, depending on contemporary circumstances, disrupting events, and social capabilities. Technological innovation, for example, provides a way to circumvent traditional meeting spaces, enhance digital interaction, and encourage file sharing, but it also exposes income and wealth inequalities, the digital divide between rich and poor, and the inability of vulnerable members of a population to access resources. Remote work eliminates commutes, improves air quality, and encourages firms to provide flexible workplaces, but it also devastates city centers. As these examples demonstrate, human societies may or may not unify in particular events; however, the events exist as the amalgamation of society's features. In the short term, belief systems, networks, and methods of organization integrate or dissipate, wax or wane, complicating the characterization of transformation. As Joel Berglund (2010) explains, "major contours or themes are made up of many small factors, and that oversimplifying those connections and factors will inevitably produce results that at best are inadequate and at worst wrong and misleading." As a result, we must acknowledge that large-scale disturbances such as global pandemics and climate change alter the existing order, leave a trail of disruption, and challenge the ability of societies to adapt. Disruptions of the existing order establish new pathways, leading to an array of potential future outcomes. But the longevity of societies, in antiquity and modern times, transcends short-term fluctuations.

Chapter thesis and organization

Large-scale disruptions, which may manifest as a series of cascading crises, impact societies, alter the existing order, and lead to a range of potential outcomes, including **resilience** or **vulnerability** (Figure 8.1). Resilience means the ability

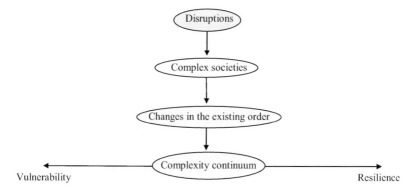

FIGURE 8.1 Disruption, change, and the complexity continuum.
Source: Author.

to absorb disturbance and retain form, processes, and structure. Vulnerability
means an inability to withstand a hostile environment. To develop these points,
the chapter discusses complex society arrangements, investment in complexity,
disruptions, a range of potential outcomes, development, and the complexity
continuum.

Complex society arrangements

In his seminal work on *The Collapse of Complex Societies*, Joseph Tainter (1988)
argues that, as societies develop, their level of complexity increases, meaning the

> size of a society, the number and distinctiveness of its parts, the variety of
> specialized social roles that it incorporates, the number of distinct social
> personalities present, and the variety of mechanisms for organizing these
> into a coherent, functioning whole.

Through the augmentation of these elements, complexity increases. **Complex
societies** are characterized by distribution and character, stability and interconnection, and evolution and differentiation.

Distribution and character

In the study of complexity, inequality refers to vertical differentiation, an unequal
distribution of resources and opportunities. Examples include wealth and income
inequality. Inequality stems from both economic factors, such as development,
exchange, and regulation, and social factors, including discrimination, oppression, and prejudice. While the forms of economic and social instability in Chapters 3–7 involve inequality, the **heterogeneity** of a human population means
the quality or state of being diverse in character or content. A heterogeneous

distribution divides occupations in an assorted and diverse manner. It is complex. The development of economic systems, sectors, and markets leads to the growth of heterogeneous occupations. In contrast, a homogeneous distribution allocates occupations equally among the population. It is not complex. As Tainter (1988) explains, "Inequality and heterogeneity are interrelated, but in part respond to different processes, and are not always positively correlated in sociopolitical evolution." For example, low levels of inequality and heterogeneity characterize early civilizations. But greater access to resources leads to higher levels of inequality, while heterogeneity remains low. Over time, as hierarchies develop, heterogeneity increases while societies become more complex. Contemporary societies demonstrate complexity with heterogeneity and varying degrees of inequality, depending on political systems and social attitudes.

Stability and interconnection

Complex societies develop systems that separate and resolve, meaning their organizing units are interdependent and relatively stable. In a historical context, a newly established nation-state may incorporate formerly independent ethnic groups or villages. An empire may include previously independent nation-states. In theory, ethnic groups, villages, or nation-states retain the potential for independence. But a shock or disruption may threaten the existing order. Depending on stability and interconnection, the decline of a complex society may create a reversion to the original form. At the end of 1991, for example, the breakup of the Soviet Union led to the reversion of nation-states that were part of the Soviet empire. In this context, the level of organization (ethnic group, village, nation-state) matters. Societies, however, increase in complexity along a continuous scale, so a discrete and stable characterization of a particular form, especially in a historical context, may be challenging to define. Modern societies include capitalist and interconnected systems but differ with respect to political representation. Nevertheless, a change in complexity within a level of organization such as a nation-state, especially in response to a major disruption, may lead to the "waxing and waning" of scale in a process of decline but not necessarily disintegration (Tainter, 1988).

Evolution and differentiation

Complex societies exist as problem-solving entities that involve structures, processes, institutions, differentiation, inequality, heterogeneity, and competition, changing as circumstances require. In a historical context, the growth of complexity exists as a movement from small, homogenous, and minimally differentiated groups with equal access to resources and homogenized occupations to large, heterogeneous, socially stratified, and internally differentiated nation-states, in which relative degrees of inequality characterize access to resources and opportunities. But modern complex societies with interconnection and access to global

networks of exchange, information, and technology are relatively recent manifestations, developing over the past few centuries and requiring augmentation,
legitimization, and reinforcement. In the process of development, investment
exists as an important factor.

Investment in complex societies

The historical record demonstrates that, in complex societies, a decline in the
existing order may occur rapidly when compared with the buildup of economic,
political, and social institutions. Tainter (1988) identifies an important reason:
diminishing returns. A common topic in the field of economics, the idea is
that, in the process of development, societies require investment in more complex public and private structures and processes to maintain social cohesion and
solve problems. But, as the structures and processes expand, they become less
efficient. Eventually, returns on investment decline. Over time, according to
Tainter, societies struggle to address new challenges, disruptions, and problems.
The potential exists for decline of the existing order and, if the process continues
without course correction, permanent damage.

Interpretations

Interpretations of a change in the existing order highlight the problem of causa-
tion between explanatory factors (initiating events) and dependent variables
(outcomes such as progress or decline). What is the true nature of change? What
is the extent to which identifiable causes, such as pandemics or economic contraction, lead to a change in the existing order? In complex societies, does economic inequality expedite the process of decline or exist as a normal feature of
the existing order? As Tainter (1988) explains, a single event may serve as the
initiating factor (Pompeii destroyed by Mount Vesuvius) or multiple economic,
political, and social causes may exist (fall of the Roman Empire). Tainter also
argues that diminishing returns (proportionately smaller benefits from additional
investment) exist as an important factor. Before presenting the model of diminishing returns, however, the features of complex adaptive systems—introduced
in Chapter 1—are relevant for the current discussion.

Complex adaptive systems

Complex adaptive systems entail dispersed interactions between multiple agents,
such as individuals, households, and businesses. The interactions lead to outcomes in interconnected processes, some in response to normal circumstances
and others in response to new circumstances. As a result, complex adaptive systems operate between the conditions of order and disorder. The extent to which
dispersed interactions, interconnected processes, and institutional characteristics create continuity and organization, order persists. But disorder may prevail.

First, inefficiency, inequality, inequity, incompetence, and instability may lead to chaos, disorganization, and disarray. Second, a disturbance may create cascading effects throughout a system. Third, both of these conditions may hold, for example, when a global pandemic ravages societies with unequal income distributions.

Diminishing returns

In the tension between order and disorder, Tainter (1988) argues that

> complex societies are more costly to maintain than simpler ones, requiring greater support levels per capita. As societies increase in complexity, more networks are created among individuals, more hierarchical controls are created to regulate these networks, more information is processed . . . there is increasing need to support specialists not directly involved in resource production.

With greater complexity, support for social systems rises. But investment exists for different purposes. The benefits of investment flow to the winners of class competition, especially elites that maintain privilege. Resources flow to those in need. With competition or integration, greater complexity exists to solve problems. The method of evaluation involves the ratio of marginal benefit to marginal cost. When favorable, investment exists as a successful strategy. When unfavorable, investment does not increase net benefits.

Model framework

In Figure 8.2, the benefits of investment in complex societies follow a characteristic curve, common in the field of economics. As institutions, processes, and structures develop, benefits initially increase (b_1 to b_2) and then reach a maximum (b_2). After this point, decline occurs gradually but then accelerates. To maintain the status quo, society must invest more resources and/or rely on technological innovation.

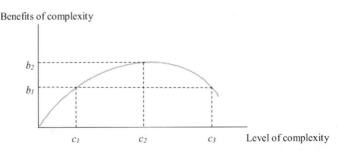

FIGURE 8.2 Diminishing returns.
Source: Tainter (1988).

Recurrent aspect of sociopolitical evolution

Tainter (1988) argues that "Diminishing returns . . . are a recurrent aspect of sociopolitical evolution, and of investment in complexity." Examples include mining precious metals during the Roman Empire, oil extraction in the Middle East, and agricultural productivity around the world. Because complexity increases systematically, interconnected elements move in the direction of progress, while others follow. For example, as agricultural systems expand, investments in bureaucracy, energy systems, and hierarchy occur. An expanded public sector must then guide additional assets, requiring its own investment in legitimization and propagation.

The struggle to counter adversities

As society invests more resources in these areas, fewer resources are available to support diversity, heterogeneity, and progressive ideals. The implication is that complexity requires an expanding resource base and/or technological innovation, each of which is subject to the law of diminishing returns. This society, in Tainter's (1988) words,

> is investing ever more heavily in a strategy that is yielding proportionately less. Excess productive capacity will at some point be used up, and accumulated surpluses allocated to current operating needs. There is, then, little or no surplus with which to counter major adversities.

The society must then address adversities such as pandemics, economic contraction, climate change, and social instability with resources out of the current operating budget. In a world of resource scarcity, this action exists to the detriment of society as a whole. "Even if the stress is successfully met," Tainter (1998) continues, "the society is weakened in the process, and made even more vulnerable to the next crisis." In the presence of diminishing returns, the emergence of an insurmountable catastrophe may simply require the passage of time. However, several forms of disruption alter complex societies.

Disruptions

Disruptions such as global pandemics and climate change destabilize societies, making it more difficult to cope with existing problems and solve new ones. But simple explanations that link disruptions to a change in the existing order are too superficial to accommodate complex adaptive systems. Two reasons exist. First, complex adaptive systems encounter disruptive forces on a regular basis. Second, they respond to disruptive forces in a variety of ways, some successful and others not. Because of these realities, disrupting forces lead to different future pathways.

Future pathways

After disruptions occur, multiple elements, including distinctive parts, specialized roles, and a variety of mechanisms, interact to generate a spectrum of future pathways. In this process, the ability of societies to innovate may exceed the ability of disruptions to weaken existing systems. But the opposite may hold true. As a result, it is important to both identify and describe disrupting forces. This approach achieves two objectives. First, it establishes a context for the era of cascading crises. Second, it provides a framework to chart a future course for society. Multiple forms of disruption—economic, environmental, external, political, and social—alter complex societies. A framework of these forces guides decision-making, policy implementation, and risk analysis.

Framework of disrupting forces

The features that define modern societies—artistic achievement, cultural endowment, economic organization, political representation, and social stratification—also define the existing order. Societies develop with these factors, advance because of them, and decline when they weaken. To establish a framework of disrupting forces, consider the range of potential outcomes. Tainter (1988),

TABLE 8.1 Disrupting forces

Category	*Disruption*	*Explanation*
Economic	Contraction	Recessionary interval
	Globalization	Global networks of connection
	Greed	Pursuit of self-interest
	Inequality	Unequal distribution of resources and opportunities
	Policy intervention	Public sector policies
	Diminishing returns	Smaller benefits from additional investment
Environmental	Climate change	Change in global temperature and weather patterns
	Degradation	Environmental deterioration
	Resource depletion	Depletion of vital resources
External	Disaster	War, famine, natural disasters, and pandemics
Political	Competition	Competition with other societies
	Folly	Lack of good judgment
	Intruders	Loss of territory or sovereignty
Social	Collective action	Working together for the common good
	Concatenations	Connected events
	Contradictions	Disputation, instability, and opposition
	Hierarchy	Social stratification

Diamond (2005), and Kemp (2019) focus on the factors that lead to the destruction of societies, at one end of the spectrum, but acknowledge that the process requires multiple generations and ruling administrations. McAnany and Yoffee (2010) focus on resilience, at the other end of the spectrum, arguing that this outcome characterizes numerous historical examples. In this chapter, the disrupting forces (Table 8.1) draw on these and other sources, including Tuchman (1984), De Vogli (2013), Diaz et al. (2019), and Bardi et al. (2019). Note that, over time, the disrupting forces may lead to positive or negative outcomes.

Economic forces

Contraction

Business cycles include four phases: expansion, peak, contraction, and trough. During contraction, economic activity declines. After the economy reaches a trough, it expands, reaches a peak, and experiences another contraction. Business cycles occur over different time frames, but the expansionary phase is usually longer than the contractionary phase.

Diminishing returns

Bardi et al. (2019) argue that diminishing returns to investment in complex systems lead to a decline in the stock of natural resources and capital. A decrease in the resource base may occur faster than the rate of economic growth. But complex systems may replenish the stocks through technological innovation and network development.

Globalization

Globalization, the interconnection of the world's people through all forms of exchange, creates networks of capital, information, migration, social interaction, technology, trade, and disease transmission. As a result, the networks facilitate both positive and negative flows. Trade, technology, information, and social connection increase consumption and production possibilities. But human transmission networks create the potential for global pandemics.

Greed

Roberto De Vogli (2013) argues that greed characterizes the global economy. Economic systems encourage profiteers, who compete for profit and wealth. But greed leads to harmful outcomes such as the Great Depression and Great Recession. In a borderless world, the globalization of greed escalates the race for profit and wealth, creating negative externalities.

Inequality

Different forms of inequality, including income and wealth inequality, restrict access to resources for marginalized members of society. Individuals with less access to education, economic opportunity, and healthcare resources have lower life expectancies and social mobility. In contrast, those with higher levels of income and wealth benefit from the existing order.

Policy intervention

Fiscal policy and monetary policy alter economic activity. Through changes in the tax code and government spending (fiscal policy) and interest rates and the money supply (monetary policy), policy intervention alters household spending and business activity. When the economy overheats, contractionary policy reduces economic activity. When the economy contracts, expansionary policy offers methods of stimulation. But a pandemic provides another reason for intervention: to stop the spread of disease. Shutdown interventions close non-essential businesses, leading to a decrease in production and employment.

Environmental forces

Climate change

Climate change includes global warming, new weather patterns, less precipitation in dry areas, more storms in wet areas, and other changes. Diamond (2005) argues that a changing climate influences the trajectory of civilizations. Over time, climate change will create harmful effects, including climate refugees, falling agricultural yields, and rising sea levels.

Degradation

Environmental degradation refers to the deterioration of the natural environment. An increase in the human population, economic growth, and disregard for environmental quality leads to degradation. Examples include air and water pollution, chemical emissions, and deforestation. An article in *Science* argues that the appropriation of nature causes the fabric of life to weaken and unravel, such as a decline in the distinctness of ecological communities, integrity of aquatic and terrestrial ecosystems, and number of species (Diaz et al., 2019).

Resource depletion

Deterioration of the resource base from climate change, environmental degradation, and human mismanagement alters future trajectories. For example, when climate change disrupts crop yields, marginal lands experience diminishing returns. This process applies pressure to more productive fields. By restricting or eliminating resources, depletion causes a decline in economic potential.

External forces

External forces of disruption include war, famine, natural disasters, and pandemics. But their risk factors vary. While a pandemic flows through global networks, famine may exist at the regional or local level. However, each factor may lead to consequential outcomes. For example, pandemics have two types of endings: the social, when fear about the spread of disease wanes, and the medical, when the number of infections and deaths approach zero. But, while infections and deaths may end, fear may remain palpable. As a result, when individuals learn to live with new conditions, a pandemic may reach its conclusion (Kolata, 2020).

Political forces

Competition

Competition between complex societies may serve as a source of disruption. Competition for economic resources encourages workers to migrate. Competition creates conflict over natural resources. Population growth leads to the cultivation of marginal land, creating disputes. In a competitive environment, capital markets create the most benefits for those at the top of the income scale, widening inequality.

Folly

Barbara Tuchman (1984), in *March of Folly*, explains that societies may experience decline not because of insurmountable obstacles but because of "wooden-headedness." Political leaders may not possess the vision and will to solve problems. By examining historical examples such as the Trojan horse, the British losing the states, and the United States losing the Vietnam war, Tuchman argues that, when mismanagement is absorbed by size or cushioned by resources, societies may maintain the status quo. But when cushions disappear, societies may not overcome folly. According to Tuchman, power corrupts. In addition, power may breed folly. In the long run, misgovernment is contrary to self-interest; however, folly causes a failure to act.

Intruders

One of the most common forms of disruption involves intruders. When invading forces occupy domestic territory or reduce the capacity to govern, societies at all levels of complexity suffer from a decline in the distinctiveness of their parts, variety of specialized roles, and types of mechanisms for organizing these into a coherent, functioning whole. A recent example is the 2022 invasion of Ukraine by Russia.

Social forces

Collective action

Acting collectively means the ability to work together to establish an objective, enhance a position, or achieve a desired outcome. The process minimizes the effects of negative flows. But when government folly, political division, economic inequality, and social instability inhibit collective action, society struggles to establish goals, solve problems, and react to changing conditions.

Concatenations

Concatenations, or connected events, describe the cascading crises in this book. The context is a lack of chance. That is, a triggering event—a global pandemic exacerbates pre-existing instabilities. When a global transmission network spreads disease, a process of contagion ravages economic, health, and social systems. The result, a decline in complexity, is manifested through lower life expectancies, economic contraction, and social instability.

Contradictions

Contradictions may be systematic, such as economies not providing equal opportunity, democratic institutions not representing those with low socioeconomic status, and social systems perpetuating the problems of exploitation and subjugation. The forms of instability described in this book, including racial injustice, domestic violence, and epistemic oppression, serve as examples. While behaviors and practices inherent in the existing order, especially by the ruling class, conceal and perpetuate these forms of instability, periods of disruption exacerbate pre-existing inequalities.

Hierarchy

Class conflict leads to antagonism. A change in the existing order may result from an uprising, when oppressed classes fight exploitation and inequality. Calls for change demand equity, representation, equal opportunity, better working conditions, and fair wages. An inability to achieve progress may lead to protest, conflict, or rebellion.

Characteristics of risk

The disrupting forces differ with respect to the characteristics of risk, introduced in Table 1.4. Depending on the characteristics, the disrupting forces may or may not have the potential for large-scale effects:

- Damage effects: measured in quantifiable units
- Delay effects: time between the triggering event and onset of damage

- Incertitude: degree of uncertainty or lack of probability
- Inequity: who bears the burden
- Mobilization: potential to address disruption in a collective manner
- Persistency: length of disruption
- Probability: likelihood of disruption as a discrete or continuous loss
- Reversibility: potential to restore the existing order
- Source: origin of disruption
- Violation: generation of cascading effects
- Ubiquity: geographic dispersion of damage

Process of identification

Using these characteristics, it is possible to identify forces that contribute to long-term and lasting effects. It is also possible to identify short-term and intermittent problems. For example, economic contraction entails damage effects in quantifiable units (job losses), a brief delay between a decrease in production and loss in employment, medium degree of incertitude, high degree of inequality, medium degree of mobilization, variable persistence, low probability of continuous loss, high potential for reversibility, clear source of disruption, medium violation in terms of the potential for cascading effects, and targeted outcomes. As a result of this risk characterization, economic contraction, a common feature of the business cycle, does not normally lead to long-term effects. For each form of disruption, the reader is encouraged to apply the risk criteria, acknowledging that risk characteristics require historical observation, contemporary judgment, and an informed forecast. But actual risk may differ from a risk assessment.

A range of potential outcomes

Modern, global societies, even powerful ones, experience disruptions with their industrial and service sectors, political structures, and social stratification. But, when confronting a series of cascading crises, societies encounter fragile, impermanent periods. They must then counter the effects of fraying domestic systems. As Tainter (1988) explains,

> Human history as a whole has been characterized by a seemingly inexorable trend toward higher levels of complexity, specialization, and sociopolitical control, processing of greater quantities of energy and information, formation of ever larger settlements, and development of more complex and capable technologies.

Yet vulnerability, especially in complex societies, exists as a recurrent theme. External disturbances, internal interruptions, and organizational miscalculations establish a position of vulnerability. In effect, times of turmoil beget interest in disruption, crisis, and aftereffects. Problems such as global pandemics, economic contractions, climate catastrophe, and social instability pressurize existing systems. This reality raises questions about the trajectory of society and the value of human civilization.

Forcing and feedback

Forcing and **feedback** characterize complex systems. Forcing, the disruption of a stable system, generates a series of enhancing or damping feedbacks. Feedbacks occur when outputs are routed back into a system as inputs. While enhancing effects amplify the forcing, damping effects decrease the amplitude of the forcing (Bardi et al., 2019). Several potential outcomes exist. Consider three. First, a local disease outbreak may alter healthcare networks and the economy, but they soon dissipate. Second, an economic recession may disrupt economic activity, trigger policy interventions, and transition to a period of expansion. Third, a global pandemic may spread through amplification channels in a process of contagion, cross a tipping point—a point at which an initial disturbance creates larger effects—and alter complex systems. In the latter case, society exists in a vulnerable state. Reducing complexity, this disturbance leads to systematic decline.

Collapse as a potential outcome

The weakening of the existing order may or may not lead to collapse. It exists as a potential but not an inevitable outcome. Jared Diamond (2005), in *Collapse: How Societies Choose to Fail or Succeed*, defines collapse as "a drastic decrease in human population size and or political/economic/social complexity, over a considerable area, for an extended time." But whether decline leads to collapse using Diamond's definition depends on interconnected factors, including the severity of the disruption, the potential for network contagion, policy responses, the potential for collective action, society's distinctiveness, variety of mechanisms, and state of vulnerability. For collapse to occur, disruptions must spread through amplification channels, reach tipping points, and destroy the capability of complex systems to adapt. The population must then disperse or change course so that the new existing order is fundamentally different from the old.

Modeling collapse

Modeling the process of collapse, Motesharrei et al. (2014) target two forms of disruption: (1) the stretching of resources due to the strain on ecological **carrying capacity**, and (2) **economic stratification**. With the first factor, carrying capacity refers to the population level that may be sustained by natural resources. With the second factor, economic stratification exists when social classes are separated, or stratified, according to economic circumstances. In model simulations, two kinds of collapse exist, due to either the scarcity of nature (depletion of natural resources) or the scarcity of labor (following an inequality-induced famine). In both scenarios, those who benefit most from the existing order—the elites—do not initially suffer, despite an impending catastrophe. Even though some identify the process of decline, advocating structural change, the elites and their supporters oppose a movement away from the existing order. According to Motesharrei et al., this mechanism helps to explain how, in historical cases of collapse, such as the Romans and Mayans, elites appeared oblivious to the catastrophic trajectory.

Regeneration as a potential outcome

From the perspective of both historical and theoretical discussions, regeneration exists as a potential outcome. Patricia McAnany and Norman Yoffee (2010), editors of *Questioning Collapse*, argue that the overriding human story is regeneration, not collapse: disturbances occur, crises exist, and conditions fluctuate, but rarely do societies "collapse in an absolute and apocalyptic sense." According to McAnany and Yoffee, a comprehensive, comparative, and historical analysis of disruptions, decision-making, and their aftereffects reveals that human resilience normally prevails. Their argument is twofold. First, some elements of complex societies may change but others remain the same. Second, the adaptive nature of complex societies means that, in longer cycles, some of these changes occur quickly but others slowly. What processes normally prevail? McAnany and Yoffee (2010) argue that the answer is "resilience, instead of collapse," because "human resilience is the rule rather than the exception . . . collapse—in the sense of the end of a social order and its people—is a rare occurrence." That is, in the presence of a global pandemic or series of cascading crises, complex systems endure through processes of adaptation and renewal.

Adaptation and renewal

The historical lessons in *Questioning Collapse*, including the medieval Norse in Greenland, Native Americans in the southwest United States, and the Aborigines in Australia, demonstrate that large-scale disruptions lead to a range of potential outcomes. While societies struggle during periods of disturbance, they may also alter their systems, implement policy measures, and adapt to changing conditions. Over the long term, as the examples in *Questioning Collapse* demonstrate, even though societies grapple with economic, environmental, external, political, and social forms of disruption, institutions, processes, and systems may remain resilient, adapting to changing circumstances.

Innovation and opportunity

Innovation creates new opportunities, decreases the price of consumer goods, and generates new markets. Innovations in energy, information, production, and transportation may avert a decline in complexity. But Tainter (1988) explains that innovation, especially the institutional variety, "is unusual in human history." Since the first industrial revolution in the late eighteenth century, however, it is common. Innovation requires investment in research and technology, a process that entails invention, innovation, and diffusion. In each stage, progress requires collaborators, network flows, and feedback loops. While advances in agriculture, production, and political organization characterize the development of complex societies, the modern era possesses the institutions and processes that incentivize innovation. Even if disruptions flow through networks of contagion, modern societies have the institutional wherewithal to address the problems.

Spectrum of possibilities

With complex societies, a spectrum of possibilities exists. Complex societies that experience disruption and diminishing returns may regenerate or decline, depending on their adaptive behavior, endowments, knowledge of threats and opportunities, methods of organization, policy measures, potential for collective action, resource allocation, and technological innovation. If these elements are robust, a society may create momentum for regeneration and progress. If these elements are weak, disruptive outcomes may lead to decline. Higher-order effects include progress, collapse, or maintenance of the status quo.

Human development

After disruption, the important question entails identification: how should we measure a change in the existing order, determining whether a society is in the process of regeneration or decline? The following method focuses on **human development**, the capacity to create opportunities, mobilize resources, and meet challenges. From society's perspective, the process of building capacity means enhancing the economic, political, and social arrangements, both public and private, that lead to human development. In contemporary and historical societies, the momentum from shared vision, collective action, and innovation strengthens this capacity. In contrast, disrupting forces weaken it. Whether the capacity expands or contracts, human development, the richness of human life, includes life expectancy, educational opportunity, and economic vitality, as specified by the United Nations (2020) in the Human Development Index (HDI).

Human development index

Published annually, the HDI exists as a way to operationalize a change in the existing order. For different countries, the HDI accomplishes three objectives. First, it establishes characteristics of human development. Second, it fosters the allocation of resources for economic, health, and social opportunities. Third, it addresses the needs of vulnerable populations. To accomplish these objectives, the index measures three dimensions of human development: (1) life expectancy,

(2) access to education (measured by expected years of schooling of children and average years of schooling of adults), and (3) standard of living (measured by gross national income [GNI], per capita, adjusted for a country's price level). The strength of the index is its ability to contrast countries over time. The weakness relates to the measurement for standard of living: an unequal income distribution skewed toward the rich increases income per capita. As a result, the measure obscures the relative depravation of individuals living at the lowest end of the income scale.

Country rankings

Top countries

The 2020 HDI demonstrates that countries at the top succeed with all three indicators (Table 8.2). Norway, for example, the country in the top position, possesses a life expectancy of 82.4 years, expected years of schooling of 18.1 years, mean years of schooling of adults of 12.1 years, and a national income per capita of $66,464.

Bottom countries

The 2020 HDI demonstrates that countries at the bottom struggle with all three indicators (Table 8.3). Niger, for example, the country in the bottom position, possesses a life expectancy of 62.4 years, expected years of schooling of 6.5 years, mean years of schooling of adults of 2.1 years, and a national income per capita of $1,201.

TABLE 8.2 Top 10 countries in the HDI (2020)

Rank	Country	HDI	Life expectancy at birth (years)	Expected years of schooling (years)	Mean years of schooling (years)	GNI per capita ($)
1	Norway	0.957	82.4	18.1	12.9	66,494
2	Ireland	0.955	82.3	18.7	12.7	68,371
3	Switzerland	0.955	83.8	16.3	13.4	69,394
4	Hong Kong	0.949	84.9	16.9	12.3	62,985
5	Iceland	0.949	83.0	19.1	12.8	54,682
6	Germany	0.947	81.3	17.0	14.2	55,314
7	Sweden	0.945	82.8	19.5	12.5	54,508
8	Australia	0.944	83.4	22.0	12.7	48,085
9	Netherlands	0.944	82.3	18.5	12.4	57,707
10	Denmark	0.940	80.9	18.9	12.6	58,662

Source: United Nations (2020), hdr.undp.org

TABLE 8.3 Bottom 10 countries in the HDI (2020)

Rank	Country	HDI	Life expectancy at birth (years)	Expected years of schooling (years)	Mean years of schooling (years)	GNI per capita ($)
180	Eretria	0.459	66.6	5.0	3.9	2,793
181	Mozambique	0.456	60.9	10.0	3.5	1,250
182	Burkina Faso	0.452	61.6	9.3	1.6	2,133
183	Sierra Leone	0.452	54.7	10.2	3.7	1,668
184	Mali	0.434	59.3	7.5	2.4	2,269
185	Burundi	0.433	61.6	11.1	3.3	754
186	South Sudan	0.433	57.9	5.3	4.8	2,003
187	Chad	0.398	54.2	7.3	2.5	1,555
188	Central African Republic	0.397	53.3	7.6	4.3	993
189	Niger	0.394	62.4	6.5	2.1	1,201

Source: United Nations (2020), hdr.undp.org

Human development over time

Because the HDI is tabulated on an annual basis, the index determines how human development changes over time. The trend depends on the country, the strength of national systems, and the ability to adapt. In a given period, some countries start with a relatively high level of human development, but others do not.

Trends

Figure 8.3 includes the two countries from the top of the HDI list, Norway and Ireland, and the two countries from the bottom, Niger and the Central African Republic. Time-series data demonstrate that, even though the countries start at different positions, a stable or upward trend exists. Periods of disruption, such as the Great Recession of 2008–2009, lead to a leveling effect. But the data demonstrate that, at the end of the disruption, a positive momentum returns.

Change in the existing order

The HDI provides a method to assess a change in the existing order. While low values in the HDI in Niger and the Central African Republic demonstrate a lack of developmental capacity, the United Nations data do not reveal an example of country-level collapse—defined to mean a large decrease in the HDI—from 1960 to 2020. With middle-income and high-income countries (such as Norway and Ireland), the United Nations data demonstrate positions of resilience, defined as an increase in HDI over time. Even though low-income countries such as Niger and the Central African Republic may not collapse, they struggle to increase

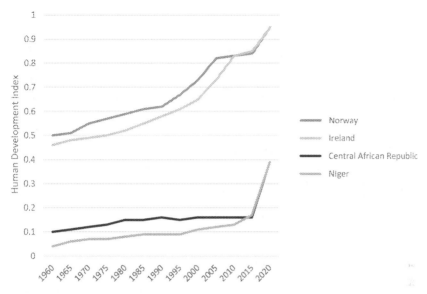

FIGURE 8.3   Change in human development.
Source: United Nations (2020), hdr.undp.org

their levels of human development. For these countries, however, the upward trend in the HDI after 2015 exists as a positive outcome.

Case study 8.1 The aftermath of pandemics

The influenza pandemic of 1918–1919 killed upwards of 50 million people, including many otherwise healthy adults. But the catastrophe left a limited impression on humanity, especially in contrast to World War I, the great war, which occurred at the same time. Perhaps the reason for this mystery relates to the century of development that separates the modern world from that period of time. Many of the victims of the influenza pandemic were born in the nineteenth century, when death from disease was common. During World War I, chemical weapons, fighter planes, and machine guns represented a technological leap in the history of disaster. In contrast, given the low life expectancy of the period, death was not unusual. In addition, the roaring twenties, characterized by growing economies and opportunity, soon followed. But, compared with the era of the influenza pandemic, contemporary society does not experience "the same cultural amnesia with Covid-19" (Couceiro, 2021). In the twenty-first century, local disease outbreaks may become national epidemics; however, prior to the coronavirus pandemic of 2020–2022, the world did not experience a global pandemic in the previous 100 years. The coronavirus pandemic's death toll, high degree of uncertainty, and economic instability altered the existing order, serving as "the source of so many genuinely new and terrifying experiences, seared into our collective memory" (Couceiro, 2021). What is the legacy of the coronavirus pandemic? First and foremost is extensive human suffering, including a large death toll. Second is the disruption of complex systems, exacerbation of pre-existing inequalities, and social division. Third is scientific breakthroughs with antibodies and vaccines. Fourth is the process of adaption. The aftermath of the coronavirus pandemic entails the vulnerabilities of complex systems, including healthcare, economies, and leadership. The aftermath also entails strengths, including innovation, scientific advancement, and human adaptation.

The complexity continuum

Complexity varies along a continuum, with resilience on one end and vulnerability on the other (Figure 8.4). Changes in the existing order signify a movement toward resilience or vulnerability. Resilience, a strong position, exists relative to the size of the society in which it occurs. But vulnerability, a weak position, may begin with a series of cascading crises and breakdown of authority, continue with economic contraction, and proceed with social instability.

Manifestations

The complexity continuum establishes potential manifestations, from small changes in the existing order to large-scale alterations, either in the direction of resilience or vulnerability. Collapse, important from a historical perspective and existing as an extreme manifestation, applies to a complete loss of economic/political/social systems. Collapse exists as a long-term process, entailing multiple generations and ruling administrations. In the short term, however, in the vulnerable end of the continuum, disrupting factors may characterize a period of decline. But the process of decline may stop short of collapse. In contrast, regeneration, characterized by the strengthening of economic/political/social systems, exists in the resilient end of the continuum. Whether regeneration leads to long-term progress depends on the potential for adaptation. With respect to these possibilities, two questions exist. For complex societies, what measures of identification establish positions along the complexity continuum? How does that position change over time? The following sections address these questions.

Vulnerability   Resilience

FIGURE 8.4 The complexity continuum.
Source: Author.

Measures of identification

The HDI establishes a country's relative position of resilience or vulnerability; however, a more comprehensive list contributes additional measures of identification:

- Economic and occupational specialization (Tainter, 1988)
- Income per capita (HDI)
- Investment (Tainter, 1988)
- Production possibilities (World Input-Output Database)
- Technological innovation (Omri, 2020)
- Resource allocation (Tainter, 1988)
- Excess deaths (*The Economist*, 2021)
- Life expectancy (HDI)
- Governance (Tuchman, 1984)
- Political representation (Hessami and da Fonseca, 2020)
- Civil liberties (Freedom House)
- Coordination and organization (Tainter, 1988)
- Expected years of schooling (HDI)
- Flow of information (Shin et al., 2020)
- Mean years of schooling (HDI)
- Stratification and social differentiation (Tainter, 1988)

Interdependence

Interdependence between the measures influences a society's position. For example, a large number of years of schooling for adults, responsive government, and a high level of income per capita establish a strong position. In addition, the measures reveal why some societies respond effectively to changing circumstances. Strong economies and government institutions provide the resources necessary to respond to disruptions.

Change in the measures of identification

The measures of identification fluctuate, alter the existing order, and create periods of regeneration and decline. Complexity correlates with higher degrees of interconnection. But, as occupations differentiate, instability persists. When specialization and income per capita increase, autonomy and self-sufficiency decline. As government expands, resource allocation becomes more inflexible. At the same time, resources in the public and private sectors flow to non-critical forms of production. In this context, "Disruptions occurring anywhere will be spread everywhere, whereas in less complex settings a society would be cushioned against disruptions by less specialization, less interlinkage among parts, and greater time delays between cause and ultimate outcomes" (Tainter, 1988). In any scale or timeframe, the factors that define the existing order do not represent universal characteristics.

Vicious cycle

Pandemics alter the measures of identification. By causing economic and social unrest, pressurizing government institutions, and weakening the existing order, pandemics lead to cascading crises. Pandemics increase excess deaths and decrease life expectancies. They push economies into shutdown, increase unemployment, and reduce labor force participation. For lower-income households, a decrease in the employment-to-population ratio perpetuates inequality. These factors create a vicious cycle (Sedik et al., 2020).

Deeper complications

Deeper complications may exist, including incompetent government, insufficient social safety nets, and political division. Disease outbreaks and social unrest create an environment where some groups fear the behaviors and attitudes of other groups, establishing the potential for instability. Economic, health, and social damage from pandemics, especially when they exacerbate inequality, exist as scarring events (Barrett et al., 2021).

Potential for transformation

Complex societies have the ability to transform. While vulnerability represents a position of weakness, changes in the existing order create new beginnings. It is, therefore, easier to identify in retrospect historical periods of decline than the unseeing eyes living at the time. In the current era, identification of a change in the existing order exists as a difficult task. Although the modern world experiences a continuous flow of disruptions, it takes time to establish a new world order. Ulrich Beck (2009), in *World at Risk*, provides a perspective: "We dramatize the decline of values, freedom, democracy, etc., so as to avoid having to acknowledge the catastrophic collapse of our own certainties about the world." As the world transforms, attitudes, beliefs, and processes change. But pandemics, economic collapse, climate catastrophe, and social instability demonstrate that vulnerable groups bear the greatest burden.

Summary

Complex societies develop the number and distinctiveness of its parts, size, variety of specialized and social roles, social personalities, and a variety of mechanisms for organizing these into a coherent whole. Investments in complexity lead to diminishing returns, in which the benefits of additional investments decline. Large-scale and interconnected disruptions, which may manifest as a series of cascading crises, alter complex societies. Disrupting forces exist in economic, environmental, external, political, and social categories. While societies struggle during periods of disruption, they also adjust their behaviors, systems, and policies, adapting to changing circumstances. Disruptions alter the existing order, leading to a range of potential outcomes, including resilience or vulnerability. One way to measure a change in the existing order is with the HDI, which includes indicators for life expectancy, education, and income per capita. Another way is to establish a comprehensive list of measures that characterize a change in the existing order.

Chapter takeaways

LO1   An alteration of the existing order means that society establishes a new reality.

LO2   Complex societies are defined as entities with distinctive parts, specialized social roles and personalities, and a variety of mechanisms for organizing these into a coherent whole.

LO3   Diminishing returns means that, as societies invest more in complex systems, additional benefits decline, making the societies more vulnerable to disruptions.

LO4   Multiple forms of disruption—economic, environmental, external, political, and social—alter complex societies.

LO5   Societies that face disruption may experience progress, regeneration, decline, collapse, or maintenance of the status quo.

LO6   Human development entails longer life expectancies, greater access to education, and higher levels of income per capita.

LO7   The complexity continuum demonstrates the potential to establish a position of resilience or vulnerability.

Key terms

Carrying capacity
Complex societies
Diminishing returns
Economic stratification
Feedback

Forcing
Heterogeneity
Human development
Resilience
Vulnerability

Questions

1   How does a change in the existing order lead to the establishment of a new reality?

2   Characterize complex societies. What is the role of inequality and heterogeneity?

3   Describe the concept of diminishing returns. How does it relate to complex societies? When a society faces a pandemic, what is the implication of diminishing returns?

4   How do disruptions reverberate throughout society?

5   How do forcings and feedback contribute to regeneration or decline?
6   For a specific country, find time-series data in the HDI. What is the trend? What factors explain the trend?
7   Facing disruption, how may a society maintain a position of resilience?
8   How do economic, health, political, and social measures define a country's existing order? How do changes in the measures signify a movement toward resilience or vulnerability?

References

Bardi, Ugo, Falsini, Sara, and Perissi, Ilaria. 2019. "Toward a general theory of societal collapse: a examination of Tainter's model of the diminishing returns of complexity." *BioPhysical Economics and Resource Quality*, 4(3): 1–12. https://doi.org/10.1007/s41247-018-0049-0.

Barrett, Philip, Chen, Sophia, and Nabar, Malhar. 2021. "Social Repercussions of Pandemics." IMF Working Paper. Washington, DC: International Monetary Fund.

Beck, Ulrich. 2009. *World at Risk*. New York: Polity.

Berglund, Joel. 2010. "Did the medieval norse society in greenland really fail?" In McAnany, Patricia and Yoffee, Norman (Eds.), *Questioning Collapse: Human Resilience, Ecological Vulnerability, and the Aftermath of Empire*. Cambridge: Cambridge University Press.

Couceiro, Cristiana. 2021. "The Great Aftermath." *The New York Times Magazine*, November 28.

De Vogli, Roberto. 2013. *Progress or Collapse: The Crises of Market Greed*. New York: Routledge.

Diamond, Jared. 2005. *Collapse: How Societies Choose to Fail or Succeed*. New York: Viking.

Diaz, Sandra, Settele, Josef, Brondizio, Eduardo…Zayas, Cynthia. 2019. "Pervasive human-driven decline in life on Earth points to the need for transformative change." *Science*, 366: 1327.

Hessami, Zohal and da Fonseca, Mariana. 2020. "Female political representation and substantive effects on policies: a literature review." *European Journal of Political Economy*, 63: 101896.

Kemp, Luke. 2019. "Are We on the Road to Civilization Collapse?" *BBC Future*, February 18.

Kolata, Gina. 2020. "How Pandemics End." *The New York Times*, May 10.

McAnany, Patricia and Yoffee, Norman. 2010. "Why we question collapse and study human resilience, ecological vulnerability, and the aftermath of empire." In McAnany, Patricia and Yoffee, Norman (Eds.), *Questioning Collapse: Human Resilience, Ecological Vulnerability, and the Aftermath of Empire*. Cambridge: Cambridge University Press.

Motesharrei, Safa, Rivas, Jorge, and Kalnay, Eugenia. 2014. "Human and nature dynamics (HANDY): modeling inequality and use of resources in the collapse or sustainability of societies." *Ecological Economics*, 101: 90–102.

Omri, Anis. 2020. "Technological innovation and sustainable development: does the stage of development matter?" *Environmental Impact Assessment Review*, 83: 106398.

Sedik, Tahsin, Xu, Rui, and Stuart, Alison. 2020. "A Vicious Cycle: How Pandemics Lead to Economic Despair and Social Unrest." IMF Working Paper. Washington, DC: International Monetary Fund.

Shin, Jaeweon, Price, Michael, Wolpert, David, Shimao, Hajime, Tracey, Brendan, and Kohler, Timothy. 2020. "Scale and information-processing threshold in Holocene social evolution." *Nature Communications*, 11: 2394.

Tainter, Joseph. 1988. *The Collapse of Complex Societies*. Cambridge: Cambridge University Press.

The Economist. 2021. "Tracking Covid-19 Deaths Across Countries." *The Economist*, December 13.

Tuchman, Barbara. 1984. *The March of Folly: From Troy to Vietnam*. New York: Ballantine Books.

United Nations. 2020. *The Next Frontier: Human Development and the Anthropocene*. New York: United Nations Development Program.

9

RESILIENCE AND VULNERABILITY

Chapter learning objectives

After reading this chapter, you will be able to:

LO1   Evaluate whether the human game is faltering.
LO2   Contrast the concepts of resilience and vulnerability.
LO3   Identify the link between intersectionality and crisis management.
LO4   Explain that agency and empowerment facilitate responsibility and self-determination.
LO5   Analyze effective governance.
LO6   Consider the potential for future resilience.

Chapter outline

Falter
Resilience and vulnerability
Intersectionality and crisis management
Agency and empowerment
Governing the commons
Building future resilience
Summary

Falter

In his book, *Falter*, Bill McKibben (2019), the environmental activist, writing before the onset of the coronavirus pandemic, argues that, because of the growing climate catastrophe, the human game, "the sum total of culture and commerce

DOI: 10.4324/9781003310075-12

and politics, of religion and sport and social life," while beautiful, complex, and deep, is "endangered" and "beginning to falter." Because human existence has no obvious ending, McKibben uses the metaphor of a game: our existence absorbs everyone's concentration. For the players, when the human game creates dignity—the state or quality of being worthy of honor or respect—it is going well. But, when dignity diminishes, society struggles. McKibben discusses the perils of climate change, citing examples of stress and degradation, including record temperatures, carbon emissions, climate refugees, rising sea levels, and drought. He highlights a study that provides a "warning to humanity" from "widespread misery and catastrophic biodiversity loss" and an inability to "shift course away from our failing trajectory" (Ripple et al., 2017).

Because of these forms of stress and degradation, McKibben argues that the world is in an early stage of transformation, characterized by technological innovation on the one hand but displacement, scarcity, and uncertainty on the other. In a warming planet, dangerous feedback loops—when some portion of the output of a system returns as an input—may occur when melting ice in the Arctic stops reflecting the Sun's rays, leading to oceanwater absorbing the Sun's heat and higher temperatures. According to McKibben (2019):

> We emit 40 gigatons of carbon dioxide annually at the moment. Our leaders express pride that we seem to be plateauing around that level, but that level is the fastest rate at any time in the last 300 million years. . . . What a large team of scientists in 2017 called a "biological annihilation" is already well under way, with half the planet's individual animals lost over the last decades and billions of local populations of animals already lost. . . . And now we are, far more rapidly than ever before in Earth's history, filling the atmosphere with the precise mix of gases that triggered the five great mass extinctions.

These factors characterize the human game, create uncertainty, and exist in the present. As *Understanding Global Crises* explains, however, climate change serves as one crisis in a series of cascading crises, along with the coronavirus pandemic, economic volatility, and social instability. These disruptions impact all areas of human civilization, threaten human dignity, and elevate McKibben's argument of faltering civilization.

Reversal of gains

During the last century, gains in health, education, and economic opportunity enhanced the indicators of human development; however, cascading crises threaten to reverse the gains. During 2020–2021, excess deaths, an increase in the number of people who die in a period relative to the average, increased in countries around the world (The Economist, 2021). During the same period, many children experienced less schooling (Engzell et al., 2021). While employment gains during the recovery interval replaced some of the lost jobs from the pandemic, more women than men quit the labor force, struggling with their schedules and the changing work environment (Kochhar and Bennett, 2021). During this period, the crises affected the most vulnerable members of society first and foremost (Kuran et al., 2020). Poverty, discrimination, and inequality exacerbated social conditions, leading to a rise in both domestic violence and racial injustice (Campbell, 2020; Wrigley-Field, 2020). A global mental health crisis emerged, characterized by anxiety, depression, and stress (Vowels et al., 2022).

By carelessness and by design

Disruptions do not guarantee a decline in complex societies. Economic and social pressures may trigger innovation, regeneration, and renewal. But population growth, globalization, and industrialization lead to a collision with the world's natural limits, increasing vulnerability and decreasing ecological **sustainability**. Here, sustainability means the ability to meet the needs of present generations without compromising the ability of future generations to meet their needs. As McKibben (2019) explains, because of **interconnection**, **consumption**, and **scale**, "We're simply so big, and moving so fast, that every decision carries enormous risk." First, the world's interconnection (relationship between agents) offers a degree of stability. But the potential of global contagion eliminates the safety of distance. Second, global consumption patterns lead to the perpetuation of extraction economies. This process degrades the natural environment. Third, the growing scale of production networks, supply chains, and markets amplifies the profit motive. Economic **incentives** then influence other systems, such as social networks. However, certain processes and achievements stabilize civilization, such as modern medicine, the capacity for learning, and collective vision. But human transmission networks, contagion, and negative flows exist as destabilizing mechanisms. Using McKibben's (2019) terminology, these latter processes are putting the human game at risk, creating leverage to both weaken existing systems and reduce living standards, "by carelessness and by design."

Chapter thesis and organization

In response to a series of cascading crises, society may move toward a greater level of resilience (strength, progress, stability) or vulnerability (weakness, regression, falter). The outcome depends on two factors: the disruptions and society's adaptive capacity. In this concluding chapter, lessons from *Understanding Global Crises* provide a method of organization: resilience and vulnerability, intersectionality and crisis management, agency and empowerment, governing the commons, and building future resilience (Table 9.1).

TABLE 9.1 Lessons from *Understanding Global Crises*

Factor	Lesson
Resilience and vulnerability	Resilience and vulnerability depend on exposure, sensitivity, and adaptive capacity
Intersectionality and crisis management	In crisis management, intersectionality serves as a guiding principle
Agency and empowerment	In social and natural systems, agency and empowerment create balance
Governing the commons	To achieve both collective and individual benefits, society must govern the commons
Building future resilience	In the presence of cascading crises, humanity has a responsibility to build future resilience

Source: Author.

Resilience and vulnerability

In Chapter 8, Figure 8.4 demonstrates the complexity continuum, along which a society may establish a position of resilience or vulnerability. The position depends on several factors, such as organizational capability, economic development, collective action, and external influences. In this context, large-scale disturbances exist as complex-system problems that entail outcome-specific responses, integrating data from multiple sources and establishing community-level solutions (Cains and Henshel, 2019). Complex-system problems include aggregation, correlation, dynamics, and scale.

Aggregation

A society's position along the complexity continuum entails the aggregation of individual attitudes, behaviors, and decisions. As a result, no single metric quantifies a system's level of resilience or vulnerability. Rather, multiple factors characterize a society's position. Because a disruption alters institutional capabilities, methods of organization, and the potential for collective action, collective responses determine whether a society withstands a disruption.

Correlation

A society's position along the complexity continuum correlates with the indicators of human development, including life expectancy, educational opportunity, and income per capita. As Chapter 8 explains, human development ranges from a high level (Norway) to a low level (Niger). Although exceptions occur, societies that establish resilient positions also excel with respect to human development. Vulnerable societies struggle with this process.

Dynamics

Along the complexity continuum, dynamic conditions persist. As a result of collective action, organizational capacity, learning, and other factors, a society's position may change. But critical thresholds may create an abrupt process: "Rather than exerting a gradual change, complex systems such as the Earth and social systems might undergo radical and abrupt shifts after crossing certain thresholds referred to as tipping points or catastrophic bifurcations" (Otto et al., 2017). Small disturbances may trigger large-scale alterations, establishing new equilibrium conditions.

Scale

A society's position along the complexity continuum is a function of scale. Because national assessments relate to macroeconomic conditions and federal government policy, they may establish a position of resilience for a country. At the same time, local assessments, which observe community capabilities, resource scarcity, and inequality, may establish a position of vulnerability.

Objectives

Aggregation, correlation, dynamics, and scale demonstrate that the objectives of society are twofold: to establish a position of resilience along the complexity continuum and strengthen that position over time. But, as *Understanding Global Crises* explains, large-scale disruptions such as pandemics, economic collapse, climate change, and social instability complicate these objectives. Strategies of **adaptation**, when societies become better suited to their environments, are important in establishing positions of resilience. To address this concept, the following sections explain resilience and vulnerability in more detail.

The concept of resilience

Resilience exists as a focus of policy and research in economic, environmental, and social contexts. Resilience refers to the ability of systems to respond to disruptions or crises. In general, resilient systems maintain or enhance their functions and objectives. Responding to disruptions, households and communities attempt to re-establish their patterns of behavior. In this context, resilience exists as a boundary concept, facilitating communication, integrating stakeholders, and establishing consensus (Baggio et al., 2015). For three reasons, a resilience framework provides a useful tool of analysis. First, it identifies potential disruptions and crises, including pandemics, economic collapse, climate change, and social instability. Second, because disruptions and crises do not exist in isolation, it identifies challenges across complex societies. As an example, climate change has "unequal contributions to the problem globally, disproportionate impacts on future generations, marginalized groups and poorer citizens (whose poverty may itself be the result of historical inequities) and asymmetries in decision-making power to determine appropriate responses" (Tanner et al., 2014). Third, a resilience framework reveals that responses to disruptions and crises are a function of multiple factors, such as institutional capacity, ideologies, inequality, planning, poverty, power relations, rights, responsibilities, risk management, and social mobilization. In the academic literature, the following two forms of resilience inform strategies of adaptation.

Social resilience

The concept of **social resilience** refers to "the buffer capacity or the ability of a system to absorb perturbations, or the magnitude of disturbance that can be absorbed before a system changes its structure by changing the variables and processes that control behavior" (Adger, 2000). Social resilience is differentiated with respect to internal and external factors. Internal factors include age, disability, ethnicity, health status, religion, and sex. External factors include cultural knowledge, education, political power, social networks, socioeconomic class, and types of assets and housing. Social resilience depends on both individual and institutional capabilities, the latter including governance, economic organization, and social behavior. Depending on legitimacy, history, and trust, institutions address risk, establish agendas, and contribute to the existing order.

Livelihood resilience

The concept of livelihood resilience places a greater emphasis on the potential for change, existing as "the capacity of all people across generations to sustain and improve their livelihood opportunities and well-being despite environmental, economic, social and political disturbances" (Tanner et al., 2014). For the most vulnerable members of society, cascading crises destabilize **livelihood networks**, the assets, capabilities, and resources that contribute to living standards. Even though livelihood resilience is a function of individual and collective action, individuals are the main agents of change. In this context, freedom exists as a human right. Governments must protect and support the process of human development; therefore, "a livelihood perspective places people at the center of the analysis, located within, rather than dominated by, ecosystems, technologies, political contexts, markets, and resource networks" (Tanner et al., 2014).

The concept of vulnerability

While vulnerability, the exposure of individuals or groups to stress from disruptions and crises, entails a loss of security, it is sensitive to institutional context. Vulnerability, which is influenced by both socioeconomic and biophysical conditions, describes the extent to which a society is susceptible to damage.

Some societies have the economic standing, institutional capacity, and social organization to address disturbances. But vulnerable societies may fail to anticipate, identify, and solve problems, because the problems are too technically difficult, possess expensive solutions, or strike with excessive speed. In this context, four important points include vulnerability as a form of distinction, result of scale, function of exposure, sensitivity, and adaptability, and threat to human well-being.

Vulnerability as a form of distinction

Vulnerability varies across physical space and socioeconomic conditions. With respect to physical space, coastal communities face a risk of sea-level rise but not inland communities. Cities located in deserts face a greater risk from drought than cities located near fresh-water sources. Societies with access to local and regional markets withstand supply shocks better than societies that rely on global networks. With respect to socioeconomic conditions, egalitarian societies address social instability better than those with histories of racism and sexism.

Vulnerability and the scale of analysis

Vulnerability depends on the scale of analysis, the relationships among and between places that reveal spatial patterns. As O'Brien et al. (2004) explain, at the national level, Norway does not experience a high degree of vulnerability to climate change; however, at the local level, a higher level of vulnerability exists. Changes in climate parameters vary, creating different levels of exposure. Although adaptation may occur, not all regions, communities, or socioeconomic groups possess the experience and resources that are necessary for adaptive responses. Communities that rely on natural resources for their economic livelihoods are more vulnerable to higher temperatures. Low-income households are more vulnerable to climate disruptions. A country may adapt, but smaller units within the country may not: "to cope with actual and potential changes . . . it will be necessary for policy makers, sectoral associations, and local institutions to acknowledge . . . vulnerabilities at the regional and local levels . . . and to address them accordingly" (O'Brien et al., 2004).

The concept of **criticality** refers to a situation in which crises disrupt complex systems, given adaptation strategies and social capabilities. This concept demonstrates that vulnerability exists to a certain degree, even within societies. For example, rising sea levels over the course of the century will place many coastal cities, including Miami, at a risk of flooding. Over time, Miami will be in a critical position. Inundation will become so severe that the city will not be able to withstand the ongoing deluge. Miami will struggle to maintain its physical systems and level of human development. During the process, vulnerable members of the community with low-income status and intersectional forms of oppression will exist in the most critical positions.

Vulnerability as a function of exposure, sensitivity, and adaptability

Three factors influence a society's degree of vulnerability. Exposure means the degree of stress on a specific unit, such as a community, hospital system, or marginalized group. Sensitivity means the degree to which a system responds to a disturbance. Adaptability refers to the capacity to adjust to forecasted or actual disruptions. Societies that create systems with low degrees of exposure, high degrees of sensitivity, and high degrees of adaptability respond effectively to disturbances (O'Brien et al., 2004).

Vulnerability as a threat to human well-being

A lack of health, safety, food security, and stability create higher levels of risk (Figure 9.1). Large-scale disruptions aggravate the health outcomes of individuals with intersectional forms of oppression. Safety is compromised in the presence of conflict, insecurity, and social breakdown. Changes in food security occur either directly through crop failures or indirectly through a decrease in supply. Displacement, resulting from housing shortages, conflict, and inequality, places housing-insecure individuals in vulnerable positions (Otto et al., 2017).

Intersectionality and crisis management

During a crisis, society should protect vulnerable individuals: they have higher levels of exposure, stress, and uncertainty. For crisis management, an intersectional perspective provides guiding principles. As Chapter 1 explains, intersectionality describes the way social forms of categorization interact to establish human identity. The intersectional theory describes how authority, power, and control impact those who experience multiple forms of oppression. Kuran et al.

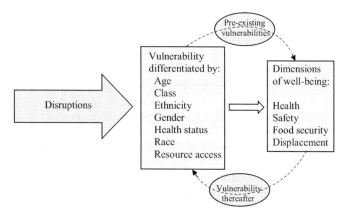

FIGURE 9.1 Threats to human well-being.
Source: Adapted from Otto et al. (2017), fig. 1, p. 1653.

(2020) argue that an intersectional framework identifies "multiple dimensions" of marginalization, economic and social risks, appropriate responses, and the needs of vulnerable members of society, including healthcare, affordable housing, and employment. During a disruption or crisis, those living on the margins of society suffer the most from racial injustice, domestic violence, and epistemic oppression.

Policy implications

The intersectional theory acknowledges that individuals experience disruptions and crises in different ways, creating the need to move beyond one-dimensional characterizations of homogenous groups, such as men and women. Rather, it acknowledges interdependent and heterogeneous elements of identity, including age, citizenship, disability, ethnicity, nationality, race, and sex. Across economic, environmental, political, and social dimensions, this framework identifies (1) differences among individuals in cultural ideologies, behavioral practices, and historical experiences; (2) outcomes of these differences; and (3) relationships within power structures. An intersectional policy framework addresses the needs of vulnerable individuals while acknowledging macro trends in employment and public health. During periods of destabilization, the intersectional framework elucidates how policy responses, such as shutdown interventions, perpetuate inequalities (Kuran et al., 2020).

Crisis management

When a disruption threatens the structures and processes of complex societies, it creates both urgency and multiple future pathways. However, it is not the disrupting event in isolation that creates specific outcomes—such as rising infections and unemployment—but in combination with adaptive capacity and intersectional realities. The coronavirus pandemic ravaged poor and immigrant populations but increased profit for high-tech companies. The shutdown interval created a period of instability for caretakers but a stable environment for those without children who could work from home. Kuran et al. (2020) emphasize that taking into account

> the intersectional perspective in relation to vulnerability and vulnerable groups means to challenge the diffuse tendency in public policy to statically categorize groups in terms of vulnerability to hazards, which neglects the differentials and fluidity of the composition within groups, in terms of vulnerability and resilience.

Because vulnerability and resilience have situational, spatial, and temporal dimensions, they exist in specific timeframes, locations, and pathways. These evolving conditions determine specific outcomes.

Agency and empowerment

The Human Development Report of the United Nations (2020), discussing the coronavirus pandemic and climate change, argues that "agency and empowerment can bring about the action we need if we are to live in balance with the planet in a fairer world." Agency refers to the capacity of individuals to act independently. Empowerment means the degree of self-determination awarded to individuals by society. Individuals with both agency and empowerment have the tools, skills, and organizational capabilities to represent their interests in informed, responsible, and thoughtful ways. Establishing agency and empowerment involves a global context and mechanisms of change.

Global context

In the current epoch, the Anthropocene, humans for the first time have the ability to shape the planet, leading to the destabilization of the climate. Even though humans have been in existence for a small fraction of the Earth's history, their imprint on the world has become the dominant cause of environmental change. Economic activity—production, distribution, and consumption—leads to carbon emissions, deforestation, extinction, and losses in biodiversity and habitat. Emerging diseases, notably Covid-19, stem from dynamic interactions between human hosts, agents of transmission, and the environment. Anthropogenic changes in food production, distribution, and consumption facilitate these interactions. Environmental degradation from the Amazonian rainforest in Brazil to the tar sands in Canada leads to damage effects. Environmental disruptions

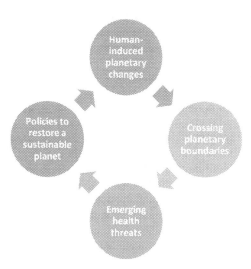

FIGURE 9.2 Policies to restore a sustainable planet.
Source: Tong et al. (2022).

in the atmosphere, oceans, and terrestrial ecosystems threaten living standards. As Tong et al. (2022) explain, the crossing of planetary boundaries creates global health threats, including the spread of novel coronaviruses; as a result, policies should restore a sustainable planet (Figure 9.2).

Mechanisms of change

In a period of destabilization, understanding global crises—Covid-19, economic collapse, climate catastrophe, and social instability—provides a glimpse of our future. The current pressure on the environment stems from economic activity and climate change, revealing fragile systems, overlapping inequalities, and uneven levels of human development. As the United Nations (2020) explains, "the challenges of planetary and societal imbalance are intertwined: they interact in a vicious circle, each making the other worse." For a sustainable future trajectory, complex societies must establish new pathways, strengthen the indicators of human development— life ex pectancy, access to education, and income per capita—and ease planetary pressures. To achieve these objectives, mechanisms of change—social norms, incentives, and nature-based solutions—create a context for agency and empowerment.

Social norms

As elements of complex societies, social norms exist as the informal rules that govern behavior; however, for vulnerable members of society, they may lead to negative outcomes. This discrepancy exists because individuals align their behavior with their peers, creating patterns of social interaction. Individuals associate these patterns with normalcy: the patterns correspond with both expectations and ideas of appropriate behavior. If everyone plays their role, social norms persist. The problem is that social norms may entail biases, discrimination, exploitation, and oppression. These negative outcomes contribute to systems of social stratification. Important elements include behavior, learning, values, and new social norms.

Behavior and learning

Through choice and cooperation, self-interest and social goals influence individual behavior. The learning process creates value for the enhancement of well-being. When behavior and learning are stable and predictable, social equilibrium occurs. However, during disruptions on the scale of Covid-19, disequilibrium may result. In response, individuals socially distance, wear masks, and avoid large gatherings. A period of disruption enhances the potential for division and instability.

Learning and values

To restore social systems, reforms should target fairness, innovation, and stewardship. To achieve these goals, social norms that inform choices on production, distribution, and consumption should establish rules to reduce social imbalances. But an important link exists between learning and values. Learning leads to both common goals and attitudes toward self-interest. That is, learning establishes values, which are guided by the public sector, private sector, and civil society.

Values and new social norms

Changing attitudes, behaviors, and preferences influence human values, leading to new social norms. But roadblocks exist. Individuals may believe they lack influence, viewing entrepreneurs or governments as the agents of change. However, individual actions lead to changes in the existing order when they are emulated and directed toward specific outcomes. When individuals initiate changes in communities, economies, and political systems, different outcomes may result. Examples include rallying for climate justice, fighting for fair wages, and protesting domestic violence. As the United Nations (2020) explains (Figure 9.3),

> through self-reinforcement, positive feedback loops, and trial and error, one or several equilibria of behavior can be reached without external intervention. By adopting new behavioral patterns, one or more individuals can shape population-level dynamics, leading to transformational change in behavior at the societal level.

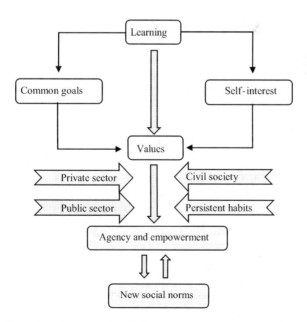

FIGURE 9.3   From learning to new social norms.
Source: Adapted from United Nations (2020), fig. 4.1, p. 135.

Incentives

One of the most important principles in economics, perhaps the most important, is that people respond to incentives. Properly crafted incentives influence human behavior. Implemented as punishment or reward, incentives reinforce social norms or initiate processes of change. Lower prices, for example, provide the incentive for an increase in quantity demanded. Farm policy provides the incentive for the production of commodity crops. Penalties from vaccination mandates provide the incentive for compliance. Incentives, in other words, help to explain patterns of consumption, production, and behavior, which are the choices of complex societies. In periods of disruption, however, incentives nudge human behavior in one direction or another. Examples include incentives for universal vaccination, altruism, and educational attainment. In the Anthropocene, properly crafted incentives also address large-scale imbalances in economic, environmental, and social systems, advancing the indicators of human development. According to the United Nations (2020), three areas of opportunity exist: finance, prices, and international collaboration.

Finance

For social transformation, mobilizing financial resources serves as an essential tool. Financial networks transfer resources from savers to borrowers. The profit motive establishes incentive for the development of new goods and services, such as cellphones, electric vehicles, and social networks. However, because capitalist systems reward profit, global networks of exchange create the economic, environmental, and social imbalances highlighted in this book, including inequality, climate change, and racial injustice. To alter the course of humanity, properly crafted financial incentives encourage equity, justice, and sustainability. Many examples exist. Investment in low-carbon power creates cleaner production processes. Fossil fuel divestment leads to sustainable investment strategies. Investment in equitable healthcare networks hedges against future pandemics. Aligning incentives with long-term objectives reduces negative outcomes during disruptive periods, channels investment, and facilitates transformation (United Nations, 2020).

Prices

In the absence of government intervention, market prices do not reflect the external cost of environmental degradation. In the absence of a pollution price, when businesses degrade the environment, the market does not value the external cost of environmental damage. As the United Nations (2020) explains, this market distortion leads to the "overuse of resources and excessive environmental degradation relative to what would occur if prices reflected those costs." Properly crafted incentives that adjust price, however, account for external cost. Carbon charges, for example, increase the price of output in industries that emit carbon. But an additional problem relates to fossil fuel **subsidies**, per-unit payments from the government. Billions of dollars' worth of annual fossil fuel subsidies complicate the movement to clean energy, which is necessary to reduce planetary imbalances. The elimination of fossil fuel subsidies reduces both carbon emissions and deaths from air pollution.

International collaboration

To address global problems such as pandemics, economic collapse, and climate change, the world requires international collaboration. In recent decades, examples include the eradication of smallpox in 1980, implementation of the Montreal protocol in 1989 to protect the ozone layer, and the economic response in 1998 to the Asian Financial Crisis. The problem is that global collaboration creates short-term costs. For individual countries, this reality may create a context for hesitation. As a model, however, the 2015 Paris agreement on climate change, adopted by 196 countries, provides a framework for global cooperation. Even though the pledges do not guarantee the goal of a reduction in carbon emissions, the agreement represents an important commitment for climate mitigation (United Nations, 2020).

Nature-based solutions

Both social norms and incentives create a context for transformational change; however, nature-based solutions—the actions to protect and manage natural and modified ecosystems—promote well-being while restoring the natural world. These actions establish agency and empowerment, while fostering innovation, stewardship, and equity (Figure 9.4).

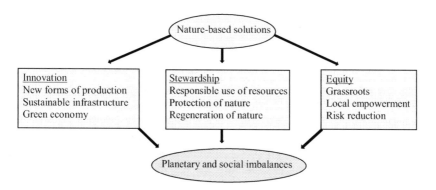

FIGURE 9.4 Nature-based solutions.

Source: Adapted from United Nations (2020), fig. 6.1, p. 185.

Nature-based solutions stem from different sources, relying on local communities and indigenous populations. These groups preserve local resources, maintain environmental flows, and support treaties. But local initiatives alter outcomes across multiple scales. The preservation of forests and grassland mitigates the effects of climate change. The restoration of green spaces in urban areas diminishes the harmful effects of urban heat waves. The maintenance of environmental quality contributes to a position of resilience. Implemented across all levels of human development, nature-based solutions benefit the global population (United Nations, 2020).

Governing the commons

The human game, as McKibben (2019) describes it, encompasses the entirety of existence, including cultural achievements, social improvements, and historical trends. But it creates institutions that extract natural resources, expand markets, and exploit marginalized individuals. The outcomes—economic insecurity, environmental degradation, and excessive levels of greenhouse gases—increase the level of vulnerability of complex societies.

Institutional capacity

Institutional capacity—the quality of leadership, personnel, and systems that produce results according to goals and objectives—puts society in a position to address large-scale problems. This capacity exists as a function of public mandates, resource allocation, and risk management. The institutions that establish capacity include governing authorities, social groups, and other forms of organization. They establish the rules and norms that characterize the existing order. In the absence of institutional capacity, society has the potential of existing in a state of peril. Large-scale problems undermine economic, environmental, and social systems.

Economic governance

In 2009, the economist Elinor Ostrom won the Nobel Prize in Economic Sciences for her analysis of economic governance, the institutions and procedures that achieve economic objectives. She focuses on problems that relate to the commons: the natural resources belonging to or affecting a community or group, including air, water, and the Earth. The implication of Ostrom's analysis is that, to achieve both collective and individual benefits, including clean air, stable water supplies, and viable natural resources, society must manage the commons in a sustainable way, balancing the needs of current generations with those of future generations. This oversight involves legal and social practices that maintain the resource base. It also requires long-term and collective perspectives, which are necessary to solve new problems.

The challenge is that preserving environmental quality requires nuanced methods of intervention under the conditions of complexity, scarcity, and uncertainty. In the presence of evolving technologies and social arrangements, effective governance requires policies, rules, and incentives that respond to these constraints.

Elinor Ostrom and her coauthors, Thomas Dietz and Paul Stern, in an article in *Science*, argue that society creates systems of effective governance by monitoring resource consumption, moderating growth in economies, and establishing social capital, the networks of relationships that facilitate effective oversight (Dietz et al., 2003). The challenge, according to the authors, is to establish these factors by using selective pressure, adaptive governance, and evolving strategies.

Selective pressure

Institutions and human behavior establish the conditions for both governance and resource allocation. A community may decide, for example, to strengthen systems of environmental oversight. But, as complex systems grow, selective pressure to establish effective forms of governance also stems from broader networks of influence. Environmental degradation exists at all levels of scale. Economic inequality has local, regional, and national implications. Social problems such as racial injustice, domestic violence, and epistemic oppression demonstrate systematic imbalances. The point is that governing capacity at all levels enables the oversight of competition, conflict, and exchange. Selective pressure shapes economic, environmental, and social outcomes, either directly or indirectly, providing a context for efforts that enhance the indicators of human development (Dietz et al., 2003).

Adaptive governance

Adaptive governance helps to fight harmful disruptions, including pandemics, climate change, and economic collapse. This process includes the provision of information, conflict resolution, and methods of compliance. First, adaptive governance requires accurate and trustworthy information about the state of society, characteristics of disrupting events, and market outcomes. While congruent in scale with shocks to the system, effective governance minimizes the impact of disinformation campaigns. Second, power asymmetries inherent in complex societies between those who benefit most from the existing order and those who do not create the potential for conflict over resources, methods of oversight, and power dynamics. The attention economy benefits those with stake in the process of technological innovation; however, it perpetuates inequality, disinformation, and instability. Adaptive governance acknowledges the diversity of human interests, perspectives, and philosophies. Third, adaptive governance establishes flexible rules, procedures, and strategies. Properly crafted incentives create a context for compliance, pollution abatement, recognition of intersectional status, and access to resources and opportunity. While different incentives exist, institutions that establish rules, procedures, and strategies exist as legitimate sources of influence (Dietz et al., 2003).

Evolving strategies

The principles and requirements of adaptive governance, according to Dietz et al. (2003), entail several links (Figure 9.5). Boundaries for individuals and groups, for example, establish methods of conflict resolution and strong institutions. Institutional variety aids conflict resolution and strategies of adaptation. Deliberation and informed discussion lead to compliance and accurate information. The point is that, to address disruptions, multiple forms of intervention exist.

Building future resilience

A period of cascading crises creates discontinuity, a time when experience and expertise fail to achieve desired results. The reason is society is in constant engagement with unpredictable conditions, including the global spread of an infectious virus, rising temperatures, economic insecurity, social conflict, fires, floods, droughts, and many other forms of instability. In the presence of discontinuity, society should build future resilience. But, in order to achieve this goal, what should society prioritize? Elinor Ostrom (2009) argues that society should maintain social-ecological systems, the "identification and analysis of relationships among multiple levels of these complex systems at different spatial and temporal scales." Chinwe Speranza et al. (2014) argue for the maintenance of human livelihoods, putting "people, their differential capabilities to cope with shocks and how to reduce poverty and improve adaptive capacity at the center of

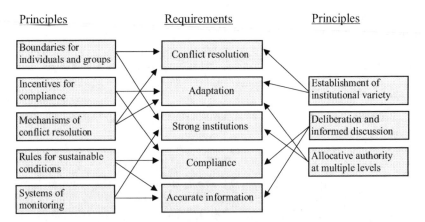

FIGURE 9.5 Principles and requirements of adaptive governance.
Source: Adapted from Dietz et al. (2003), fig. 3, p. 1910.

analysis." Bob Doppelt (2017) argues that, to address the existential problem of climate change, society should use the "best available science" and do "what is possible to prepare human-built infrastructure, agriculture, water systems, and other natural systems" for the onslaught of global damage effects. Kevin Grove (2018) explains that, in the presence of "precariousness and insecurity in all facets of everyday life," society should "harness the dynamic and creative force of change . . . and use this force to create sustainable social and ecological systems in a complex and indeterminate world." Bill McKibben (2019) concludes that society should move "toward a smaller-scale world less obsessed with efficiency." In McKibben's view, economic growth in the past provided more benefit than cost but the opposite now holds true. As a result, McKibben explains that the current period of instability requires maturity, balance, and appropriate scale. That is, society should balance the needs of present generations with those of future generations, minimize damage effects, and pursue sustainable future pathways. Together, the arguments demonstrate the need to address threats from multiple sources, not just climate change and pandemics, but also economic volatility and social instability. To build future resilience, complex societies should establish a context for capacity, learning, organization, strategic intervention, and transformation.

Capacity, learning, and organization

Buffer capacity, self-organization, and capacity for learning serve as important elements of future resilience (Speranza et al., 2014). Buffer capacity refers to the level of change that a system may absorb while maintaining function, identity, and structure. Self-organization entails the norms, values, and organization that exist without control or constraint. The capacity for learning connotes strategies of adaptive management that turn experiences into action. These dimensions, when properly crafted, maintain or increase agency and empowerment, improving the "understanding of people's adaptive capacities," their "differential capabilities to cope with shocks," and "people's livelihoods" (Speranza et al., 2014). As an application, the dimensions may be decomposed into proxy indicators (Table 9.2). For example, societies with buffer capacity, self-organization, and capacity for learning, exemplified through social capital, cooperation, and collective vision, are in a position to build future resilience. In contrast, societies that experience division, myopic behavior, and a lack of collective action struggle to build future resilience, existing in a vulnerable state.

Strategic intervention

To fight climate change, Grove (2018) asks how to design "a more resilient city that can survive and thrive despite looming climate change impacts, even if survival involves transforming into a new and unrecognizable . . . form?" Because of the identification of damage effects, Grove's question applies to economic,

TABLE 9.2 Dimensions of livelihood resilience

Dimensions	Indicators	Meaning
Buffer capacity	Financial capital	Wealth used to maintain or enhance economic activity
	Human capital	Attributes useful for production
	Natural capital	World's stock of natural resources
	Physical capital	Tangible resource inputs for production
	Social capital	Networks of relationships among individuals in society
Self-organization	Cooperation	Process of working together
	Institutions	Significant practices, relationships, and organizations
	Networks	Systems of interconnection
	Opportunity	Circumstances that establish possibilities
	Reliance	Dependence or trust in something or someone
Capacity for learning	Collective vision	Actions or objectives shared by members of society
	Commitment	State or quality of being dedicated to a cause or activity
	Feedbacks	System outputs routed back as inputs
	Knowledge	Awareness or familiarity gained through experience

Source: Speranza et al. (2014).

environmental, and social contexts. With pressure from different forms of instability, how should complex societies design resilient systems? As an example, Grove (2018) discusses water pumps in Miami, which run around the clock in a (futile) effort to keep rising sea waters out, embodying a "kind of promise, the promise of resilience . . . that even as sea levels rise, even as the world changes around us, we might be able to adjust to these new realities." In contrast to water pumps in Miami, however, resilient practices reconfigure human actions and the objectives of human systems. They exist as a desire to synthesize forms of knowledge and new practices in order to establish collaborative and flexible solutions to "contextually specific problems of complexity" (Grove, 2018). While water pumps in Miami do not exist as long-term solutions, strategic intervention, according to Grove (2018):

> emerges out of the process of collaborating on complex problems that have no straightforward, predictable solutions. It is contingent on pragmatically identifying ways that different (and limited) forms of scientific knowledge can advance collaborate efforts to (re)define specific problems that need to be addressed, develop different kinds of interventions, monitor the effects

of these interventions to ensure they are having desirable impacts on social and ecological systems, and adjust these interventions as needed.

Methods of strategic intervention emphasize the need to establish, in the living labs of communities, innovative solutions to problems of complexity.

Transformation

Doppelt (2017) argues that transformational resilience involves decisions that enhance, rather than diminish, human well-being. The process requires the development of support networks. The framework not only focuses on threats to the natural environment and climate but also applies to economic, health, and social pressures, because damages are:

> indisputably traumatic and exceedingly stressful, producing significant effects on the human mind and body. Without landmark efforts to enhance the capacity of individuals and groups for Transformational Resilience, the traumas and toxic stresses—meaning persistent overwhelming stresses—will generate unprecedented levels of anxiety, depression, post-traumatic stress disorder . . . and other mental health problems.

What elements characterize transformational resilience? The first is presencing, an act that involves the identification of personal strength and skills, internal mental resources, support networks, external resources, and connections to other

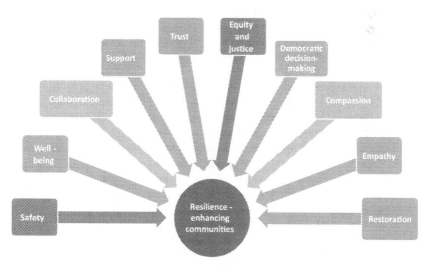

FIGURE 9.6 Traits of resilience-enhancing communities.
Source: Adapted from Doppelt (2017), fig. 11.2, p. 275.

support networks. The objective is to establish a context for new insights and understanding. But it also entails the strengthening of human connections. The second element of transformational resilience is purposing, the act of identifying sources of meaning, establishing core values, and creating hope for new possibilities. The idea of transformational resilience, according to Doppelt, is to use the elements of presencing and purposing to establish the traits of resilience-enhancing communities (Figure 9.6).

Case study 9.1 Group behavior during periods of disruption

Humans have an unending desire to establish groups. They also invest in community. Working cooperatively, expanding economic opportunity, and establishing methods of organization exist as fundamental elements of complex societies. Within communities, however, humans identify with group membership, such as political parties, fanbases, and clubs. The idea is that groups are meaningful and productive, creating space for connection, sharing, and common values. In periods of disruption, however, humans increase their volunteer efforts, whether working at Covid clinics, homeless shelters, or foodbanks, providing aid to those in need. In communities, the efforts provide volunteers with a sense of belonging. Over time, as bonds strengthen, the rewards for volunteer efforts grow. The outcomes exist as both moral and valuable choices. However, during periods of crisis, volunteer actions seem more powerful, especially in the presence of rising deaths, layoffs, domestic violence, and racial injustice. As Jon Mooallem (2021) explains,

> Disasters are disruptions, cleaving ordinary life open into vacuums of uncertainty. Falling into a community with others—particularly a community acting with purpose—helps people regain certainty and agency. In a catastrophe, taking action can feel restorative, euphoric even, and seems to help sustain survivors' mental health. . . . If people feel that their group is in danger or that a disaster has occurred, they will take action in innovative ways.

The problem is that interaction during periods of disruption increases the potential for misinformation. Some individuals may not have the experience, insight, or motivation to question the propagation of false news. In systems of interconnection, antisocial behavior may prevail. Some groups undermine the process of sustaining complex societies, but others use collective action to help their neighbors.

Summary

Cascading crises, including pandemics, economic collapse, climate catastrophe, and social instability, create harmful outcomes: rising infections, deaths, unemployment, climate effects, racial injustice, domestic violence, and misinformation. Whether or not the human game, the sum of commerce, culture, politics, religion, and social life, is faltering depends on perspective, lasting effects from the cascading crises, and human responses. The complexity continuum demonstrates that society may establish a position of resilience or vulnerability. But disruptions and crises threaten to place society in a more vulnerable position. The intersectionality theory describes how race, ethnicity, gender, and other forms of social categorization interact to establish human identity. Systems of authority, power, and control impact those who experience multiple forms of oppression. The existence of intersectional outcomes influences strategies of crisis management. Agency and empowerment create methods to live in balance with the planet in a fairer world. While institutional capacity puts a society in position to address large-scale problems, effective governance requires selective pressure, adaptation, and evolving strategies. Building future resilience means strengthening existing institutions, processes, and behaviors so a society maintains its structure and form in the presence of large-scale crises.

Chapter takeaways

LO1   Despite the complexity of the human game, it is beginning to falter.
LO2   On the complexity continuum, disruptions move society toward vulnerability.
LO3   An intersectional perspective provides principles to help vulnerable members of society.
LO4   Agency and empowerment increase the potential for balance and equity.
LO5   Effective governance requires selective pressure, adaptation, and evolving strategies.
LO6   To build future resilience, complex societies should establish a context for capacity, learning, and organization, methods of strategic intervention, and processes of transformation.

Key terms

Adaptation	Livelihood networks
Consumption	Scale
Criticality	Social resilience
Incentives	Subsidies
Interconnection	Sustainability

Questions

1 In what sense, if any, is the human game beginning to falter?
2 How do exposure, sensitivity, and adaptability determine the degree of vulnerability?
3 Using an intersectional perspective, why are individuals experiencing multiple forms of oppression more vulnerable during crises?
4 How may mechanisms of change (social norms, incentives, and nature-based solutions) create a context for agency and empowerment?
5 Why do selective forms of pressure, methods of adaptive governance, and evolving strategies serve as effective elements of governance?
6 How do buffer capacity, self-organization, and capacity for learning help societies build future resilience?
7 In building future resilience, what is the role of strategic intervention? Identify examples.
8 What is transformational resilience? What elements characterize the process?

References

Adger, W. Neil. 2000. "Social and ecological resilience: are they related?" *Progress in Human Geography*, 24(3): 347–364.

Baggio, Jacopo, Brown, Katrina, and Hellebrandt, Denis. 2015. "Boundary object or bridging concept? A citation network analysis of resilience." *Ecology and Society*, 20(2): 1–11.

Cains, Mariana and Henshel, Diane. 2019. "Community as an equal partner for region-based climate change vulnerability, risk, and resilience assessments." *Current Opinion in Environmental Sustainability*, 39: 24–30.

Campbell, Andrew. 2020. "An increasing risk of family violence during the Covid-19 pandemic: Strengthening community collaborations to save lives." *Forensic Science International: Reports*, 2: 100089.

Dietz, Thomas, Ostrom, Elinor, and Stern, Paul. 2003. "The struggle to govern the commons." *Science*, 302: 1907–1912.

Doppelt, Bob. 2017. *Transformational Resilience*. New York: Routledge.

The Economist. 2021. "Tracking Covid-19 Excess Deaths across Countries." *The Economist*, December 30.

Engzell, Per, Frey, Arun, and Verhagen, Mark. 2021. "Learning loss due to school closures during the Covid-19 pandemic." *PNAS*, 118(17): e2022376118.

Grove, Kevin. 2018. *Resilience*. New York: Routledge.

Kochhar, Rakesh and Bennett, Jesse. 2021. "U.S. Labor Market Inches Back from the Covid-19 Shock, but Recovery Is Far from Complete." *Pew Research Center*, April 14.

Kuran, Christian, Morsut, Claudia, and Kruke, Bjorn…2020. "Vulnerability and vulnerable groups from an intersectionality perspective." *International Journal of Disaster Risk Reduction*, 50: 101826.

McKibben, Bill. 2019. *Falter: Has the Human Game Begun to Play Itself Out?* New York: Henry Holt and Company.

Mooallem, Jon. 2021. "Is Life Better When We're Together?" *The New York Times*, December 26.

O'Brien, Karen, Sygna, Linda, and Haugen, Jan. 2004. "Vulnerable or resilient? A multi-scale assessment of climate impacts and vulnerability in Norway." *Climatic Change*, 64(1–2): 193–225.

Ostrom, Elinor. 2009. "A general framework for analyzing sustainability of social-ecological systems." *Science*, 325(24): 419–422.

Otto, Ilona, Reckien, Diana, Reyer Christopher…2017. "Social vulnerability to climate change: a review of concepts and evidence." *Regional Environmental Change*, 17: 1651–1662.

Ripple, William, Wolf, Christopher, Newsome, Thomas…2017. "World Scientists' Warning to Humanity: A Second Notice." *BioScience*, 67(12): 1026–1028.

Speranza, Chinwe, Wiesmann, Urs, and Rist, Stephan. 2014. "An indicator framework for assessing livelihood resilience in the context of social-ecological dynamics." *Global Environmental Change*, 28: 109–119.

Tanner, Thomas, Lewis, David, Wrathall, David…2014. "Livelihood resilience in the face of climate change." *Nature Climate Change*, 1: 23–26.

Tong, Shilu, Bambrick, Hilary, Beggs, Paul…2022. "Current and future threats to human health in the Anthropocene." *Environment International*, 158: 106892.

United Nations. 2020. *The Next Frontier: Human Development and the Anthropocene*. New York: United Nations Development Program.

Vowels, Laura, Carnelley, Katherine, and Stanton, Sarah. 2022. "Attachment anxiety predicts worse mental health outcomes during Covid-19: evidence from two studies." *Personality and Individual Differences*, 185: 111256.

Wrigley-Field, Elizabeth. 2020. "US racial inequality may be as deadly as Covid-19." *PNAS*, 117(36): 21854–21856.

INDEX

Printed in the United States
by Baker & Taylor Publisher Services